Principles of Evolution

Principles of Evolution

Solomon Stevens

R CALLISTO REFERENCE

www.callistoreference.com

Callisto Reference,
118-35 Queens Blvd., Suite 400,
Forest Hills, NY 11375, USA

Visit us on the World Wide Web at:
www.callistoreference.com

ISBN: 978-1-64116-234-0 (Hardback)

Cataloging-in-Publication Data

Principles of evolution / Solomon Stevens.
   p. cm.
Includes bibliographical references and index.
ISBN 978-1-64116-234-0
1. Evolution (Biology). 2. Biology. I. Stevens, Solomon.
QH366.2 .P75 2019
576.8--dc23

Table of Contents

Preface

Evolution is a natural process that drives a change in the heritable characteristics of populations over successive generations. Such changes are mediated by genes as they are transmitted from one parent to offspring. Evolution is at the core of the emergence of biodiversity at each stage of biological organization such as at species, organisms and molecular levels. Genetic variations in a population are introduced due to mutations and genetic recombination. Some of these characteristics become common or rare within a specific population due to evolutionary processes of natural selection and genetic drift. The impact of evolution is seen in terms of change in adaptability, cooperation among species, coevolution, speciation and extinction. This book is a valuable compilation of topics, ranging from the basic to the most complex theories and principles in the field of evolution. The topics included herein are of the utmost significance and bound to provide incredible insights to readers. This textbook is an essential guide for both academicians and those who wish to pursue this discipline further.

To facilitate a deeper understanding of the contents of this book a short introduction of every chapter is written below:

Chapter 1- Any natural change in the heritable traits of a biological population over successive generations is termed as evolution. This is an introductory chapter, which will elucidate all the significant aspects of evolution, such as molecular evolution, emergent evolution, Darwin's theory of evolution and devolution.

Chapter 2- Evolutionary biology is a branch of biology, which studies the forces of evolution that are responsible for the emergence of biological diversity on Earth. This chapter has been carefully written to provide an easy understanding of the varied facets of evolutionary biology, such as evolutionary dynamics, evolutionary developmental biology, evolution of cells and evolutionary physiology.

Chapter 3- Natural selection is the preferential survival and reproduction of specific individuals based on differences in their heritable traits. The aim of this chapter is to explore the fundamentals of natural selection, such as heritable traits, sexual selection, stabilizing selection, directional selection, negative selection, gene selection, microevolution and macroevolution, among others.

Chapter 4- Speciation is an evolutionary process, which is responsible for the evolution of populations into distinct species. This chapter closely examines the key concepts of speciation, such as ecological speciation, allopatric speciation, peripatric speciation, parapatric speciation, cospeciation, despeciation, etc. for a holistic understanding of the subject.

Chapter 5- A permanent alteration in the nucleotide sequence of a genome of an organism is called mutation. Adaptation is a dynamic evolutionary process, which fits organisms into their surroundings, and enhances their evolutionary fitness. All the diverse principles of mutation and adaptation have been carefully analyzed in this chapter, such as mutationism, neutral mutations, suppressor mutation, point mutation, dynamic mutation, frameshift mutation, etc.

I owe the completion of this book to the never-ending support of my family, who supported me throughout the project.

Solomon Stevens

Chapter 1

Understanding Evolution

Any natural change in the heritable traits of a biological population over successive generations is termed as evolution. This is an introductory chapter, which will elucidate all the significant aspects of evolution, such as molecular evolution, emergent evolution, Darwin's theory of evolution and devolution.

Charles Darwin, father of the theory of evolution by natural selection

Broadly defined, biological evolution is any heritable change in a population of organisms over time. Changes may be slight or large, but must be passed on to the next generation (or many generations) and must involve populations, not individuals.

Similarly, the term may be presented in terms of allele frequency (with an "allele" being an alternative form of a gene, such as different alleles code for different eye colors): "Evolution can be precisely defined as any change in the frequency of alleles within a gene pool from one generation to the next". Both a slight change (as in pesticide resistance in a strain of bacteria) and a large change (as in the development of major new designs such as feathered wings, or even the present diversity of life from simple prokaryotes) qualify as evolution.

However, "evolution" commonly is used more narrowly to refer to the specific theory that all organisms have descended from common ancestors, also known as the "theory of descent with modification," or to refer to one explanation for the process by which change occurs, the "theory of modification through natural selection." The term also is used with reference to a comprehensive theory that includes both the non-causal pattern of descent with modification and the causal mechanism of natural selection.

Evolution is a central concept in biology. Geneticist T. Dobzhansky (1973) has stated, "Nothing in biology makes sense, except in the light of evolution," and biologist Ernst Mayr (2001) has stated, "Evolution is the most profound and powerful idea to have been conceived in the last two centuries."

Nonetheless, the concepts of evolution have often engendered controversy during the past two centuries, particularly from Christians, whose traditional views have been challenged both by the long time period of evolution and by the purposeless, materialistic mechanism inherent in having natural selection be the creative force. Modern Christian viewpoints range from rejecting both descent with modification (the pattern) and the mechanism of natural selection (the process), to accepting descent with modification but not the theory of natural selection, to those claiming natural selection as God's way of creating things.

The development of modern theories of evolution began with the introduction of the concept of natural selection in a joint 1858 paper by Charles Darwin and Alfred Russel Wallace, and the publication of Darwin's 1859 book, The Origin of Species. Darwin and Wallace proposed that evolution occurs because a heritable trait that increases an individual's chance of successfully reproducing will become more common, by inheritance, from one generation to the next, and likewise a heritable trait that decreases an individual's chance of reproducing will become rarer. In the 1930s, scientists combined Darwinian natural selection with the re-discovered theory of Mendelian heredity to create the modern synthesis, which is the prevailing paradigm of evolutionary theory.

Alfred Russel Wallace

Evolutionary Theory

As broadly and commonly defined in the scientific community, the term evolution connotes heritable changes in populations of organisms over time, or changes in the frequencies of alleles over time. A popular definition along these lines is that offered by Douglas J. Futuyma (1986) in Evolutionary Biology: "Biological evolution is change in the properties of populations of organisms that transcend the lifetime of a single individual. The changes in populations that are considered evolutionary are those that are inheritable via the genetic material from one generation to another." In this sense, the term does not specify any overall pattern of change through the ages, nor the process whereby change occurs (although the term is also employed in such a manner).

However, there are two very important and popular evolutionary theories that address the pattern and process of evolution: "theory of descent with modification" and "theory of natural selection," respectively, as well as other concepts in evolutionary theory that deal with speciation and the rate of evolution.

Theory of Descent with Modification

The "theory of descent with modification" is the major kinematic theory that deals with the pattern of evolution—that is, it treats non-causal relations between ancestral and descendant species, orders, phyla, and so forth. The theory of descent with modification, also called the "theory of common descent," essentially postulates that all organisms have descended from common ancestors by a continuous process of branching. In other words, narrowly defined, all life evolved from one kind of organism or from a few simple kinds, and each species arose in a single geographic location from another species that preceded it in time. Each group of organisms shares a common ancestor. In the broadest sense of the terminology, the theory of descent with modification simply states that more recent forms result from modification of earlier forms.

One of the major contributions of Charles Darwin was to marshal substantial evidence for the theory of descent with modification, particularly in his book, Origin of Species. Among the evidences that evolutionists use to document the "pattern of evolution" are the fossil record, the distribution patterns of existing species, methods of dating fossils, and comparison of homologous structures.

Theory of Natural Selection

The second major evolutionary theory is the "theory of modification through natural selection," also known as the "theory of natural selection." This is a dynamic theory that involves mechanisms and causal relationships. The theory of natural selection is one explanation offered for how evolution might have occurred; in other words, the "process" by which evolution took place to arrive at the pattern.

The term natural selection may be defined as the mechanism whereby biological individuals that are endowed with favorable or deleterious traits reproduce more or less than other individuals that do not possess such traits. Natural selection generally is defined independently of whether or not there is actually an effect on the gene-frequency of a population. That is, it is limited to the selection process itself, whereby individuals in a population experience differential survival and reproduction based on a particular phenotypic variation(s).

The theory of evolution by natural selection is the comprehensive proposal involving both heritable genetic variations in a population and the mechanism of natural selection that acts on these variations, such that individuals with greater fitness are more likely to contribute offspring to the next generation, while individuals with lesser fitness are more likely to die early or fail to reproduce. As a result, genotypes with greater fitness become more abundant in the next generation, while genotypes with a lesser fitness become rarer. This theory encompasses both minor changes in gene frequency in populations, brought about by the creative force of natural selection, and major evolutionary changes brought about through natural selection, such as the origin of new designs. For Darwin, however, the term natural selection generally was used synonymously with evolution by natural selection.

In the theory of natural selection as currently conceived, there is both a chance component and a non-random component. Genetic variation is seen as developing randomly, by chance, such as through mutations or genetic recombination. Mayr (2002) states that the production of genetic variation "is almost exclusively a chance phenomena." In every generation, new mutations and recombinations arise spontaneously, producing a new spectrum of phenotypes for natural selection—a non-random selective force (Mayr 2002)—to act upon. However, Mayr (2002) also notes that chance plays an important role even in "the process of the elimination of less fit individuals," and particularly during periods of mass extinction. Thus, chance (stochastic processes, randomness) also plays a major role in the theory of natural selection.

According to the theory of natural selection, natural selection is the directing or creative force of evolution. Natural selection is considered far more than just a minor force for weeding out unfit organisms. Even Paley and other natural theologians accepted natural selection, albeit as a mechanism for removing unfit organisms, rather than as a directive force for creating new species and new designs.

Concrete evidence for the theory of modification by natural selection is limited to microevolution—that is, evolution at or below the level of species. The evidence that natural selection directs changes on the macroevolutionary level—such as the major transitions between higher taxa and the origination of new designs—necessarily involves extrapolation from these evidences on the microevolutionary level. The validity of making such extrapolations has recently been challenged by some prominent evolutionists.

The theory of natural selection received a much more contentious response than did the theory of descent with modification. One of Darwin's chief purposes in publishing the Origin of Species was to show that natural selection had been the chief agent of the changes presented in the theory of descent with modification. While the theory of descent with modification was accepted by the scientific community soon after its introduction, the theory of natural selection took until the mid-1900s to be accepted. However, even today, this theory remains controversial, with detractors in both the scientific and religious communities.

Speciation and Extinction

The concepts of speciation and extinction are important to any understanding of evolutionary theory.

Speciation is the term that refers to creation of new and distinct biological species by branching off from the ancestral population. Various mechanisms have been presented whereby a single evolutionary lineage splits into two or more genetically independent lineages. For example, allopatric speciation is held to occur in populations that become isolated geographically, such as by habitat fragmentation or migration. Sympatric speciation is held to occur when new species emerge in the same geographic area. Ernst Mayr's peripatric speciation is a proposal for a type of speciation that exists in between the extremes of allopatry and sympatry, where zones of differentiating species abut but do not overlap.

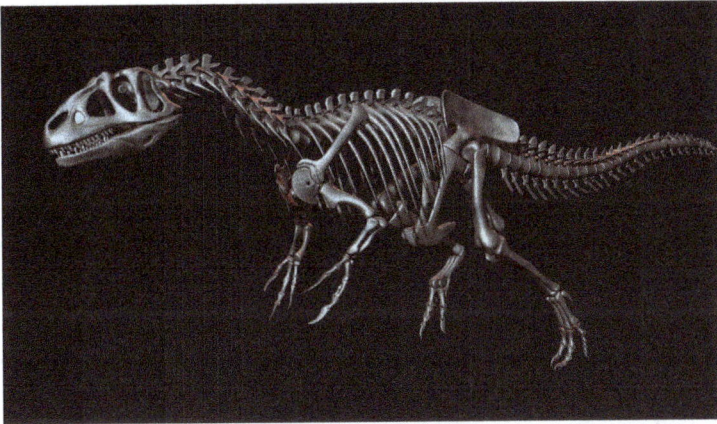

An allosaurus skeleton

Extinction is the disappearance of species (i.e. gene pools). The moment of extinction generally occurs at the death of the last individual of that species. Extinction is not an unusual event in geological time. The Permian-Triassic extinction event was the Earth's most severe extinction event, rendering extinct 90 percent of all marine species and 70 percent of terrestrial vertebrate species. In the Cretaceous-Tertiary extinction event, many forms of life perished (including approximately 50 percent of all genera), the most often mentioned among them being the extinction of the dinosaurs.

One of the unheralded laws of evolutionary theory is that macroevolutionary changes are irreversible—lineages do not return to their ancestral form, even when they return to the ancestral way of life.

Rate of Evolution

The concept of gradualism has often been linked with evolutionary thought. Gradualism is a view of descent with modification as proceeding by means of slow accumulation of very small changes, with the evolving population passing through all the intermediate stages—sort of a "march of frequency distributions" through time.

Darwin himself insisted that evolution was entirely gradual. Indeed, he stated in the Origin of Species:

- "As natural selection acts solely by accumulating slight, successive, favourable variations, it can produce no great or sudden modifications; it can act only by very short and slow steps."

- Nature "can never take a leap, but must advance by the shortest and slowest steps."

- "If it could be demonstrated that any complex organ existed, which could not possibly have been formed by numerous, successive, slight modifications, my theory would absolutely break down."

The Darwinian and Neo-Darwinian emphasis on gradualism has been subject to re-examination on several levels: the levels of major evolutionary trends, origin of new designs, and models of speciation.

1. Punctuated equilibrium

A common misconception about evolution is that the development of new species generally requires millions of years. Indeed, the gradualist view that speciation involved a slow, steady, progressive transformation of an ancestral population into a new species has dominated much of evolutionary thought from the time of Darwin. Such a transformation was commonly viewed as involving large numbers of individuals ("usually the entire ancestral population"), being "even and slow," and occurring "over all or a large part of the ancestral species' geographic range" (Eldredge & Gould 1972). This concept was applied to the development of a new species by either phyletic evolution (where the descendant species arises by the transformation of the entire ancestral population) or by speciation (where the descendant species branches off from the ancestral population).

However, paleontologists now recognize that the fossil record does not generally yield the expected sequence of slightly altered intermediary forms, but instead the sudden appearance of species, and long periods when species do not change much.

The theory of punctuated equilibrium ascribes that the fossil record accurately reflects evolutionary change. That is, it posits that macroevolutionary patterns of species are typically ones of morphological stability during their existence (stasis), and that most evolutionary change is concentrated in events of speciation—with the origin of a new species usually occurring during geologically short periods of time when the long-term stasis of a population is punctuated by this rare and rapid speciation event. The sudden transitions between species are sometimes measured on the order of hundreds or thousands of years relative to their millions of years of existence. Although the theory of punctuated equilibrium originally generated a lot of controversy, it is now viewed highly favorably in the scientific community, and has even become a part of recent textbook orthodoxy.

Note that the theory of punctuated equilibrium merely addresses the pattern of evolution and is not tied to any one mode of speciation. Although occurring in a brief period of time, the species formation can go through all the stages, or can proceed by leaps. It is even neutral with respect to natural selection.

2. Punctuated origin of new designs

According to the gradualist viewpoint, the origin of novel features, such as feathers in birds and jaws in fish, can be explained as having arisen from numerous, tiny, imperceptible steps, with each step being advantageous and developed by natural selection. Darwin's proposed such a resolution for the origin of the vertebrate eye.

However, there are some structures for which it is difficult to conceive how such structures could be useful in incipient stages, and thus have selective advantage. One way in which evolutionary theory has dealt with such criticisms is the concept of "preadaptation," proposing that the intermediate stage may perform useful functions different from the final stage. Incipient feathers may have been used for retaining body warmth or catching insects, for example, prior to the development of a fully functional wing.

Another solution for origin of new designs, which is gaining renewed attention among evolutionists, is that the full sequence of intermediate forms may not have existed at all, and instead key features may have developed by rapid transitions, discontinuously. This view of a punctuational origin of key features arose because of:

- The persistent problem of the lack of fossil evidence for intermediate stages between major designs, with transitions between major groups being characteristically abrupt;
- The inability to conceive of functional intermediates in select cases.

In the later case, prominent evolutionist Stephen Jay Gould (1980b) cites the fur-lined pouches of pocket gophers and the maxillary bone of the upper jaw of certain genera of boid snakes being split into front and rear halves: "How can a jawbone be half broken? What good is an incipient groove or furrow on the outside? Did such hypothetical ancestors run about three-legged while holding a few scraps of food in an imperfect crease with their fourth leg?"

The concept of punctuational origin is not necessarily opposed to natural selection as the creative force. For example, the rapid transition could be the product of a very small genetic change, even one mutation occurring by chance in a key gene, which is then acted upon by natural selection. However, the concept of a punctuational origin of new designs (as with punctuational equilibrium), is also viewed favorably by those advocating divine creation, due to the alignment of this view with the concept of discontinuous variation being the product of divine input, with natural selection simply the weeding out of previous, less well-adapted forms.

3. Punctuational models of speciation

Punctuational models of speciation are being advanced in contrast with what is sometimes labeled the "allopatric orthodoxy". Allopatric orthodoxy is a process of species origin involving geographic isolation, whereby a population completely separates geographically from a large parental population and develops gradually into a new species by natural selection until their differences are so great that reproductive isolation ensues. Reproductive isolation is therefore a secondary byproduct of geographic isolation, with the process involving gradual allelic substitution. Contrasted with this view are recent punctuational models for speciation, which postulate that reproductive isolation can rise rapidly, not through gradual selection, but without selective significance. In such models, reproductive isolation originates before adaptive, phenotypic differences are acquired. Selection does not play a creative role in initiating speciation, nor in the definitive aspect of reproductive isolation, although it is usually postulated as the important factor in building subsequent adaptation. One example of this is polyploidy, where there is a multiplication of the number of chromosomes beyond the normal diploid number. Another model is chromosomal speciation, involving large changes in chromosomes due to various genetic accidents.

Darwinism and Neo-Darwinism

Darwinism is a term generally synonymous with the theory of natural selection. Harvard evolutionist Stephen Jay Gould (1982) maintains: "Although 'Darwinism' has often been equated with evolution itself in popular literature, the term should be restricted to the body of thought allied with Darwin's own theory of mechanism [natural selection]." Although the term has been used in various ways depending on who is using it and the time period (Mayr 1991), Gould nonetheless finds a general agreement in the scientific community that "Darwinism should be restricted to the world view encompassed by the theory of natural selection itself."

The term neo-darwinism is a very different concept. It is considered synonymous with the term "modern synthesis" or "modern evolutionary synthesis." The modern synthesis is the most significant, overall development in evolutionary thought since the time of Darwin, and is the prevailing paradigm of evolutionary biology. The modern synthesis melded the two major theories of classical Darwinism (theory of descent with modification and the theory of natural selection) with the rediscovered Mendelian genetics, recasting Darwin's ideas in terms of changes in allele frequency.

In essence, advances in genetics pioneered by Gregor Mendel led to a sophisticated concept of the basis of variation and the mechanisms of inheritance. Gregor Mendel proposed a gene-based theory of inheritance, describing the elements responsible for heritable traits as the fundamental units now called genes and laying out a mathematical framework for the segregation and inheritance of variants of a gene, which are now referred to as alleles. Later research identified the molecule DNA as the genetic material through which traits are passed from parent to offspring, and identified genes as discrete elements within DNA. Though largely maintained within organisms, DNA is both variable across individuals and subject to a process of change or mutation.

According to the modern synthesis, the ultimate source of all genetic variation is mutations. They are permanent, transmissible changes to the genetic material (usually DNA or RNA) of a cell, and can be caused by "copying errors" in the genetic material during cell division and by exposure to radiation, chemicals, or viruses.

In addition to passing genetic material from parent to offspring, nearly all organisms employ sexual reproduction to exchange genetic material. This, combined with meiotic recombination, allows genetic variation to be propagated through an interbreeding population.

According to the modern synthesis, natural selection acts on the genes, through their expression (phenotypes). Natural selection can be subdivided into two categories:

- Ecological selection occurs when organisms that survive and reproduce increase the frequency of their genes in the gene pool over those that do not survive.

- Sexual selection occurs when organisms that are more attractive to the opposite sex because of their features reproduce more and thus increase the frequency of those features in the gene pool.

Through the process of natural selection, species become better adapted to their environments. Note that, whereas mutations (and genetic drift) are random, natural selection is not, as it preferentially selects for different mutations based on differential fitness.

In recent years, there have been many challenges to the modern synthesis, to the point where Bowler (1988), a historian of evolutionary thought, states; "In the last decade or so it has become obvious that there is no longer a universal consensus in favor of the synthetic theory even within the ranks of working biologists." Gould (1980a) likewise notes "that theory, as a general proposition is effectively dead." These challenges include models of punctuational change, the theory of "neutralism," and selection at levels above the individual. What some historians and philosophers of evolutionary thought see as challenges to the modern synthesis, others see as either erroneous theories or as theories that can be included within the umbrella of the modern synthesis.

Evidences of Evolution

For the broad concept of evolution ("any heritable change in a population of organisms over time"), evidences of evolution are readily apparent. Evidences include observed changes in domestic crops (creating a variety of corn with greater resistance to disease), bacterial strains (development of strains with resistance to antibiotics), laboratory animals (structural changes in fruit flies), and flora and fauna in the wild (color change in particular populations of peppered moths and polyploidy in plants).

Generally, however, the "evidences of evolution" being presented by scientists or textbook authors are for either

(1) the theory of descent with modification;

(2) a comprehensive concept including both the theory of descent with modification and the theory of natural selection.

In actuality, most of these evidences that have been catalogued are for the theory of descent with modification.

Evidences for the Theory of Descent with Modification

In the Origin of Species, Darwin marshaled many evidences for the theory of descent with modification, within such areas as paleontology, biogeography, morphology, and embryology. Many of these areas continue to provide the most convincing proofs of descent with modification even today. Supplementing these areas, are molecular evidences.

It is noteworthy that some of the best support for the theory of descent with modification comes from the observation of imperfections of nature, rather than perfect adaptations. As noted by Gould:

> "All of the classical arguments for evolution are fundamentally arguments for imperfections that reflect history. They fit the pattern of observing that the leg of Reptile B is not the best for walking, because it evolved from Fish A. In other words, why would a rat run, a bat fly, a porpoise swim and a man type all with the same structures utilizing the same bones unless inherited from a common ancestor?"

Fossil Record

Fossil evidence of prehistoric organisms has been found all over the Earth. Fossils are traces of once living organisms. Fossilization on an organism is an uncommon occurrence, usually requiring hard parts (like bone) and death where sediments or volcanic ash may be deposited. Fossil evidence of organisms without hard body parts, such as shell, bone, teeth, and wood stems, is sparse, but exists in the form of ancient microfossils and the fossilization of ancient burrows and a few soft-bodied organisms. Some insects have been preserved in resin. The age of fossils can often be deduced from the geologic context in which they are found (the strata); and their age also can be determined with radiometric dating.

The comparison of fossils of extinct organisms in older geological strata with fossils found in more recent strata or with living organisms is considered strong evidence of descent with modification. Fossils found in more recent strata are often very similar to, or indistinguishable from living species, whereas the older the fossils the more different they are from living organisms or recent fossils. In addition, fossil evidence reveals that species of greater complexity have appeared on the earth over time, beginning in the Precambrian era some 600 millions of years ago with the first eukaryotes. The fossil records support the view that there is orderly progression in which each stage emerges from, or builds upon, preceding stages.

One of the problems with fossil evidence is the general lack of gradually sequenced intermediary forms. There are some fossil lineages that appear quite well-represented, such as from therapsid reptiles to the mammals, and between what is considered land-living ancestors of the whales and their ocean-living descendants. The transition from an ancestral horse (Eohippus) and the modern horse (Equus) is also significant, and Archaeopteryx has been postulated as fitting the gap between reptiles and birds. But generally, paleontologists do not find a steady change from ancestral forms to descendant forms, but rather discontinuities, or gaps in most every phyletic series. This has been explained both by the incompleteness of the fossil record and by proposals of speciation that involve short periods of time, rather than millions of years. (Notably, there are also gaps between living organisms, with a lack of intermediaries between whales and terrestrial mammals, between reptiles and birds, and between flowering plants and their closest relatives.) Archaeopteryx has recently come under criticism as a transitional fossil between reptiles and birds.

The fact that the fossil evidence supports the view that species tend to remain stable throughout their existence and that new species appear suddenly is not problematic for the theory of descent with modification, but only with Darwin's concept of gradualism.

Morphological Evidence

The study of comparative anatomy also yields evidence for the theory of descent with modification. For one, there are structures in diverse species that have similar internal organization yet perform different functions. Vertebrate limbs are a common example of such homologous structures. Bat wings, for example, are very similar to human hands. Also similar are the forelimbs of the penguin, the porpoise, the rat, and the alligator. In addition, these features derive from the same structures in the embryo stage. As queried earlier, "why would a rat run, a bat fly, a porpoise swim and a man type" all with limbs using the same bone structure if not coming from a common ancestor, since these are surely not the most ideal structures for each use.

Likewise, a structure may exist with little or no purpose in one organism, yet the same structure has a clear purpose in other species. These features are called vestigial organs or vestigial characters. The human wisdom teeth and appendix are common examples. Likewise, some snakes have pelvic bones and limb bones, and some blind salamanders and blind cave fish have eyes. Such features would be the prediction of the theory of descent with modification, suggesting that they share a common ancestry with organisms that have the same structure, but which is functional.

For the point of view of classification, it can be observed that various species exhibit a sense of "relatedness," such as various catlike mammals can be put in the same family (Felidae), dog-like mammals in the same family (Canidae), and bears in the same family (Ursidae), and so forth, and then these and other similar mammals can be combined into the same order (Carnivora). This sense of relatedness, from external features, fits the expectations of the theory of descent with modification.

Phylogeny, the study of the ancestry (pattern and history) of organisms, yields a phylogenetic tree to show such relatedness (or a cladogram in other taxonomic disciplines).

Embryology

A common evidence for evolution is the assertion that the embryos of related animals are often quite similar to each other, often much more similar than the adult forms. For example, it is held that the development of the human embryo is compatible to comparable stages of other kinds of vertebrates (fish, salamander, tortoise, chicken, pig, cow, and rabbit). Furthermore, mammals such as cows and rabbits are more similar in embryological development than with alligators. Often, the drawings of early vertebrate embryos by Ernst Haeckel are offered as proof.

It has further been asserted that features, such as the gill pouches in the mammalian embryo resemble those of fish, are most readily explained as being remnants from the ancestral fish, which were not eliminated because they are embryonic "organizers" for the next step of development.

Wells (2000) has criticized embryological evidence on several points. For one, it is now known that Ernst Haeckel exaggerated the similarities of vertebrate embryos at the midpoint of embryological development, and omitted the earlier embryological stages when differences were more pronounced. Also, embryological development in some frog species looks very similar to that of birds, rather than other frog species. Remarkably, even as revered an evolutionist as Ernst Mayr, in his 2001 text What Evolution Is, used Haeckel drawings from 1870, which he knew were faked, noting "Haeckel had fraudulently substituted dog embryos for the human ones, but they were so similar to humans that these (if available) would have made the same point."

Biogeography

The geographic distribution of plants and animals offers another commonly cited evidence for evolution (common descent). The fauna on Australia, with its large marsupials, is very different from that of the other continents. The fauna on Africa and South America are very different, but the fauna of Europe and North America, which were connected more recently, are similar. There are few mammals on oceanic islands. These findings support the theory of descent with modification, which holds that the present distribution of flora and fauna would be related to their common origins and subsequent distribution. The longer the separation of continents, such as with Australia's long isolation, the greater the expected divergence is.

Renowned evolutionist Mayr contends that "the facts of biogeography posed some of the most insoluble dilemmas for the creationists and were eventually used by Darwin as his most convincing evidence in favor of evolution."

Molecular Evidence

Evidence for common descent may be found in traits shared between all living organisms. In Darwin's day, the evidence of shared traits was based solely on visible observation of morphologic similarities, such as the fact that all birds—even those which do not fly—have wings. Today, the theory of common descent is supported by genetic similarities. For example, every living cell makes use of nucleic acids as its genetic material, and uses the same twenty amino acids as the building blocks for proteins. All organisms use the same genetic code (with some extremely rare and minor deviations) to translate nucleic acid sequences into proteins. The universality of these traits strongly suggests common ancestry, because the selection of these traits seems somewhat arbitrary.

Similarly, the metabolism of very different organisms is based on the same biochemistry. For example, the protein cytochrome c, which is needed for aerobic respiration, is universally shared in aerobic organisms, suggesting a common ancestor that used this protein. There are also variations in the amino acid sequence of cytochrome c, with the more similar molecules found in organisms that appear more related (monkeys and cattle) than between those that seem less related (monkeys and fish). The cytochrome c of chimpanzees is the same as that of humans, but very different from bread mold. Similar results have been found with blood proteins.

Other uniformity is seen in the universality of mitosis in all cellular organisms, the similarity of meiosis in all sexually reproducing organisms, the use of ATP by all organisms for energy transfer, and the fact that almost all plants use the same chlorophyll molecule for photosynthesis.

The closer that organisms appear to be related, the more similar are their respective genetic sequences. That is, comparison of the genetic sequence of organisms reveals that phylogenetically close organisms have a higher degree of sequence similarity than organisms that are phylogenetically distant. For example, neutral human DNA sequences are approximately 1.2 percent divergent (based on substitutions) from those of their nearest genetic relative, the chimpanzee, 1.6 percent from gorillas, and 6.6 percent from baboons. Sequence comparison is considered a measure robust enough to be used to correct erroneous assumptions in the phylogenetic tree in instances where other evidence is scarce.

Comparative studies also show that some basic genes of higher organisms are shared with homologous genes in bacteria.

Evidences for the Theory of Natural Selection

Concrete evidence for the theory of modification by natural selection is limited to the microevolutionary level—that is, events and processes at or below the level of species. As examples of such evidences, plant and animal breeders use artificial selection to produce different varieties of plants and strains of fish. Natural selection is seen in the changes of the shade of gray of populations of peppered moths (Biston betularia) observed in England.

Another example involves the hawthorn fly, Rhagoletis pomonella. Different populations of hawthorn fly feed on different fruits. A new population spontaneously emerged in North America in the nineteenth century sometime after apples, a non-native species, were introduced. The apple-feeding population normally feeds only on apples and not on the historically preferred fruit of hawthorns. Likewise the current hawthorn feeding population does not normally feed on apples. A current area of scientific research is the investigation of whether or not the apple-feeding race may further evolve into a new species. Some evidence, such as the fact that six out of thirteen alozyme loci are different, that hawthorn flies mature later in the season, and take longer to mature than apple flies, and that there is little evidence of interbreeding (researchers have documented a 4 to 6 percent hybridization rate) suggests this possibility.

The evidence that natural selection directs the major transitions between species and originates new designs (macroevolution) involves extrapolation from these evidences on the microevolutionary level. That is, it is inferred that if moths can change their color in 50 years, then new designs or entire new genera can originate over millions of years. If geneticists see population changes for fruit flies in laboratory bottles, then given eons of time, birds can be built from reptiles and fish with jaws from jawless ancestors.

However, at question has always been the sufficiency of extrapolation to the macroevolutionary level. As Mayr, "from Darwin's day to the present, there has been a heated controversy over whether macroevolution is nothing but an unbroken continuation of microevolution, as Darwin and his followers have claimed, or rather is disconnected from microevolution."

Teaching of Evidences

Textbook authors have often confused the dialogue on evolution by treating the term as if it signified one unified whole—not only descent with modification, but also the specific Darwinian and neo-Darwinian theories regarding natural selection, gradualism, speciation, and so forth. Certain textbook authors, in particular, have exacerbated this terminological confusion by lumping "evidences of evolution" into a section placed immediately after a comprehensive presentation on Darwin's overall theory—thereby creating the misleading impression that the evidences are supporting all components of Darwin's theory, including natural selection. In reality, the confirming information is invariably limited to the phenomenon of evolution having occurred (descent from a common ancestor or change of gene frequencies in populations), or perhaps including evidence of natural selection within populations.

Evolution as Fact and Theory

"Evolution" has been referred to both as a "fact" and as a "theory."

In scientific terminology, a theory is a model of the world (or some portion of it) from which falsifiable hypotheses can be generated and tested through controlled experiments, or be verified through empirical observation. "Facts" are parts of the world, or claims about the world, that are real or true regardless of what people think. Facts, as data or things that are done or exist, are parts of theories—they are things, or relationships between things, that theories take for granted in order to make predictions, or that theories predict. For example, it is a "fact" that an apple dropped on earth will fall towards the center of the planet in a straight line, and the "theory" that explains it is the current theory of gravitation.

In common usage, people use the word "theory" to signify "conjecture," "speculation," or "opinion." In this popular sense, "theories" are opposed to "facts." Thus, it is not uncommon for those opposed to evolution to state that it is just a theory, not a fact, implying that it is mere speculation. But for scientists, "theory" and "fact" do not stand in opposition, but rather exist in a reciprocal relationship.

Scientists sometimes refer to evolution as both a "fact" and a "theory."

In the broader usage of the term, calling evolution a "fact" references the confidence that scientists have that populations of organisms can change over time. In this sense, evolution occurs whenever a new strain of bacterium evolves that is resistant to antibodies that had been lethal to prior strains. Many evolutionists also call evolution a "fact" when they are referring to the theory of descent with modification, because of the substantial evidences that they perceive as having been marshaled for this theory. In this later sense, Mayr opines: "It is now actually misleading to refer to evolution as a theory, considering the massive evidence that has been discovered over the past 140 years documenting its existence. Evolution is no longer a theory, it is simply a fact."

When "evolution" is referred to as a theory by evolutionists, the reference is generally to an explanation for why and how evolution occur.

Modern Alternative Mechanisms and Views

Symbiogenesis

Symbiogenesis is evolutionary change initiated by a long-term symbiosis of dissimilar organisms. Margulis and Sagan hold that random mutation is greatly overemphasized as the source of hereditary variation in standard Neo-Darwinistic doctrine. Rather, they maintain, the major source of transmitted variation actually comes from the acquisition of genomes—in other words, entire sets of genes, in the form of whole organisms, are acquired and incorporated by other organisms. This long-term biological fusion of organisms, beginning as symbiosis, is held to be the agent of species evolution.

For example, lichens are a composite organism composed of a fungus and a photosynthetic partner (usually either green algae or cyanobacteria, but in some cases yellow-green algae, brown algae, or both green algae and cyanobacteria). These intertwined organisms act as a unit that is distinct from its component parts. Lichens are considered to have arisen by symbiogenesis, involving acquisitions of cyanobacterial or algal genomes.

Another example is the photosynthetic animals or plant-animal hybrids in the form of slugs (shell-less mollusks) that have green algae in their tissues (such as Elysia viridis). These slugs are always green, never need to eat throughout their adult life, and are "permanently and discontinuously different from their gray, algae-eating ancestors". This is held to be another example of a symbiosis that lead to symbiogenesis.

Yet another example is cattle, which are able to digest cellulose in grass because of microbial symbionts in their rumen. Cattle cannot survive without such an association. Other examples of evolution resulting through merger of dissimilar organisms include associations of modern (scleractinian) coral and dinomastigotes (such as Gymnodinium microadriaticum) and the formation of new species and genera of flowering plants when when the leaves of these plants integrated a bacterial genome.

The formation of eukaryotes is postulated to have occurred through a symbiotic relationship between prokaryotes, a theory called endosymbiosis. According to this theory, mitochondria, chloroplasts, flagella, and even the cell nucleus would have arisen from prokaryote bacteria that gave up their independence for the protective and nutritive environment within a host organism.

Margulis and Sagan state that the formation of new species by inheritance of acquired microbes is best documented in protists. They conclude that "details abound that support the concept that all visible organisms, plants, animals, and fungi evolved by "body fusion.""

More Complex Tree of Life

The conventional paradigm of the theory of descent with modification presumes that the history of life maps as the "tree of life," a tree beginning with the trunk as one universal common ancestor and then progressively branching, with modern species at the twig ends. However, that clean and simple pattern is being called into question due to discoveries being made by sequencing genomes of specific organisms. Instead of being simple at its base, the tree of life is looking considerably more complex. At the level of single cells, before the emergence of multicellular organisms, the genomic signs point not to a single line of development, but rather to a bush or a network as diverse microbes at times exchange their genetic material, especially through the process of lateral gene transfer.

Other complicating factors are proposed based on the relatively sudden appearance of phyla during the Cambrian explosion and on evidence that animals may have originated more than once and in different places at different times.

Non-random Variation

The current paradigm of the theory of natural selection is that the process has a major stochastic (random) element, with heritable variation arising through chance, and then being acted upon by the largely non-random force of natural selection made manifest as various species compete for limited resources. An alternative view is that the introduced variation is non-random.

In particular, various theistic perspectives see directed variation, from a Supreme Being, as the creative force of evolution. Natural selection, rather than being the creative force of evolution, may be variously viewed as a force for advancement of the new variation or may be considered largely inconsequential. Some role may also be accorded differential selection, such as mass extinctions. This view sees the evolutionary process as progressive, non-materialistic, and purposeful.

Neither of these contrasted worldviews—random variation and the purposeless, non-progressive role of natural selection, or purposeful, progressive variation—are conclusively proved or unproved by scientific methodology, and both are theoretically possible.

Evolution of Life

The appearance of life on earth is not a part of biological evolution.

Not much is known about the earliest developments in life. However, all existing organisms share certain traits, including cellular structure and genetic code. Most scientists interpret this to mean all existing organisms share a common ancestor that had already developed the most fundamental cellular processes. There is no scientific consensus on the relationship of the three domains of life (Archea, Bacteria, Eukaryota) or the origin of life.

Pre-Cambrian stromatolites in the Siyeh Formation, Glacier National Park. *Nature* arguing that formations such as this possess 3.5 billion-year-old fossilized algae microbes. If true, they would be the earliest known life on earth

The emergence of oxygenic photosynthesis (around 3 billion years ago) and the subsequent emergence of an oxygen-rich, non-reducing atmosphere can be traced through the formation of banded iron deposits, and later red beds of iron oxides. This was a necessary prerequisite for the development of aerobic cellular respiration, believed to have emerged around 2 billion years ago.

In the last billion years, simple multicellular plants and animals began to appear in the oceans. Soon after the emergence of the first animals, the Cambrian explosion (a period of unrivaled and remarkable, but brief, organismal diversity documented in the fossils found at the Burgess Shale) saw the creation of all the major body plans, or phyla, of modern animals. About 500 million years ago, plants and fungi colonized the land, and were soon followed by arthropods and other animals, leading to the development of the land ecosystems of today.

Utilizing the fossil record, scientists have constructed geological timetables, or geological time scales to offer a picture of the history of life on earth, organized by presenting the type of plant and animal life according to the time of appearance (often listed in terms of era, period, epoch, and years). This timetable, for example, locates the first bacteria and the first algae in the Precambrian era, over 1 billion years ago, the first marine invertebrates in the Cambrian period of the Paleozoic era (some 580 million years ago), early mammals in the Triassic period of the Mesozoic era, the first flowering plants in the Cretaceous period of the Mesozoic era, and the development of early hominids in the Pliocene epoch of the Tertiary period of the Cenozoic era, and so forth.

One of the great puzzles in biology is the sudden appearance of most body plans of animals during the early Cambrian period and why there have been no major new structural types in the subsequent 500 million years.

Scientists also strive to show lineages, from ancestral to descendant organisms. There are numerous evidences that are used in constructing this more defined history of life, with the best known being the fossil record, but also utilizing the comparative anatomy of present-day plants and animals. By comparing the anatomies of both modern and extinct species, biologists attempt to reconstruct the lineages of those species. Transitional fossils have been proposed to picture continuity between two different lineages. For instance, the connection between dinosaurs and birds has been proposed by way of so-called "transitional" species such as Archaeopteryx.

The development of genetics also has allowed biologists to investigate the genetic record of the history of life as well. Although we cannot obtain the DNA sequences of most extinct species, the degree of similarity and difference among modern species allows geneticists to reconstruct lineages. It is from genetic comparisons that claims such as the 98 to 99 percent similarity between humans and chimpanzees come from, for instance.

Other evidence used to demonstrate evolutionary lineages includes the geographical distribution of species. For instance, monotremes and most marsupials are found only in Australia, postulating that their common ancestor with placental mammals lived before the submerging of the ancient land bridge between Australia and Asia.

Scientists correlate all of the above evidence—drawn from paleontology, anatomy, genetics, and geography—with other information about the history of the earth. For instance, paleoclimatology attests to periodic ice ages during which the climate was much cooler; and these are found to match up with the spread of species such as the woolly mammoth that are better equipped to deal with cold.

Evolution and Religion

A satirical image of Charles Darwin as an ape, reflects part of the social controversy over whether humans and apes share a common lineage

Since the publication of the The Origin of Species in 1859, the concept of evolution has engendered controversy, particularly from religious leaders. Popular writings often tend to create an artificial dichotomy—either belief in a Creator is correct or evolution is correct: evolution and religion (specifically creation by a Supreme Being or God) are presented as if mutually exclusive alternatives. Thus, many religious adherents reject evolution out of hand, not wishing to reject God.

Nevertheless, religious viewpoints are varied with respect to evolution. Some faith communities, such as "young-earth creationists" stand in opposition to both the theory of descent with modification and the theory of natural selection. Holding strictly to the letter of Genesis, they hold that the Earth is only 6,000 years old, that God created all the plants and animals in the first week of creation, and that the fossil record is actually artifacts from before the Flood.

Other believers accept the pattern observed in nature (theory of descent with modification) but not the process. They hold that God as Creator had a hand in his creations at many stages along the way from bacterium to human, imparting his design and his image. This is what is classically called creationism, or more narrowly "old-earth creationism," since it accepts the scientific account of the gradual development of life on earth over four billion years. They critique the young earth position by citing the verse, "with the Lord one day is as a thousand years, and a thousand years as one day".

Still others accept natural selection as the causal agent of large-scale change. This latter view fits that of evolutionary geneticist Theodosius Dobzhansky: "It is wrong to hold creation and evolution as mutually exclusive alternatives. I am a creationist and an evolutionist. Evolution is God's or Nature's, method of creation." Theologically, this would be a Deist position, since once God set up natural selection, it would have carried on autonomously without any activity on God's part. It might be termed "theistic evolution" but certainly not "creation."

By itself, the theory of descent with modification poses little difficulty to most religious adherents, since it is neutral with respect to the process. The mechanism that gives rise to the pattern could occur by natural selection or it could occur by the directive force of a supreme being. In 1859, most scientists and laymen believed that the biotic world was constant. The massive evidence that Darwin presented was so convincing that within a few years every biologist became an evolutionist, believing that the world was the product of a continuing process of change. For most biologists today, the view that evolution takes place—that there is a systematic change in populations—is taken as fact.

Adherents of scientific creationism, and in particular young-earth creationists, do oppose the theory of descent with modification, but they represent only a small body of those individuals that do believe in a creation by a supreme being.

Classical creationists are likewise opposed to evolution, despite having a belief system that allows descent with modification and change in gene frequencies in populations.

Mainly, they are opposed to the specific Darwinian theory of evolution by natural selection, which has three radical components that are particularly troublesome:

(1) purposelessness,

(2) philosophical materialism,

(3) lack of being progressive.

Natural selection is purposeless, requiring no input from a higher Power; it does not require God or God's purposes as an explanation for the seeming harmony in the world. Thus natural selection is opposed to creation as an active process by which God acts to mold life to his purposes. Religious believers who palpably experience God acting in their personal lives find it hard to accept that God also does not act to develop his creation.

Natural selection is materialistic, holding that matter is the main reality of existence and that mental and spiritual phenomena, including thought, will, and feeling, can be explained in terms of matter, as its byproducts. Many religious believers understand God to have created human beings for the expressed purpose of embodying reason and spirituality, by which they can know God and manifest a divine nature. A theory such as evolution, which holds that mind and spirit are mere byproducts of a materialistic process, cannot square with belief in the supremacy of mind and spirit as the highest aspects of creation.

Evolution by natural selection is not progressive from lower to higher, but just an adaptation to local environments; it could form a man with his superior brain or a parasite, but no one could say which is higher or lower. Humans are granted no special status. The view that human beings are evolved, not as a designed end-result but as if by accident, is squarely at odds with many religious interpretations.

Belief in creation by a higher power is linked with some notion of design. The designs of the creation begin in the mind of God, who forms creatures according to these designs. This is what the Gospel of John teaches by the statement: "In the beginning was the Word"—and Jewish, Muslim and Hindu scriptures have similar concepts. If there is an evolutionary process, there should be the input of God's design along the way, directing the process. In this view, protozoa cannot just evolve by a purposeless process into mammals. The creation of higher-order beings should require the investment of God's labor and thought. The development of sophisticated new designs via such a "purposeless" process as natural selection has been compared to having a hurricane assemble a 747 airplane from just the parts.

In recent years, the intelligent design (ID) movement has gained momentum in the United States. ID essentially holds that it is possible to infer from empirical evidence that some features of the natural world are best explained by an intelligent agent. This movement seeks to present in educational institutions a scientific critique of evolutionary theory and offer the possibility of living organisms being designed. Technically it is not considered a religious perspective according to many of its advocates, since it presents its views without reference to whom or what that designer may be.

Emergence of Evolutionary Thought

The idea of biological evolution has existed since ancient times, notably among Hellenists such as Epicurus and Anaximander, but the modern theory was not established until the eighteenth and nineteenth centuries, by scientists such as Jean-Baptiste Lamarck and Charles Darwin. While transmutation of species was accepted by a sizeable number of scientists before 1859, it was the publication of Charles Darwin's The Origin of Species that provided the mechanism of natural selection as the means by which evolutionary change occurs. Darwin was motivated to publish his work after receiving a letter from Alfred Russel Wallace, in which Wallace revealed his own concept of natural selection.

Darwin's theory could not explain the source of variation in traits within a species, and Darwin's proposal of a hereditary mechanism (pangenesis) was not compelling to most biologists. It was not until the late nineteenth and early twentieth centuries that these mechanisms were established.

When Gregor Mendel's work regarding the nature of inheritance in the late nineteenth century was "rediscovered," it led to a storm of conflict between Mendelians (Charles Benedict Davenport) and biometricians (Walter Frank Raphael Weldon and Karl Pearson), who insisted that the great majority of traits important to evolution must show continuous variation that was not explainable by Mendelian analysis. Eventually, the two models were reconciled and merged, primarily through the work of the biologist and statistician R.A. Fisher. This combined approach, applying a rigorous statistical model to Mendel's theories of inheritance via genes, became known in the 1930s and 1940s as the modern evolutionary synthesis.

In the 1940s, Oswald Avery, Colin McCleod, and Maclyn McCarty definitively identified deoxyribonucleic acid (DNA) as the "transforming principle" responsible for transmitting genetic information. In 1953, Francis Crick and James Watson published their famous paper on the structure of DNA, based on the research of Rosalind Franklin and Maurice Wilkins. These developments ignited the era of molecular biology and transformed the understanding of evolution into a molecular process: the mutation of segments of DNA.

George C. Williams' 1966 Adaptation and Natural Selection: A Critique of Some Current Evolutionary Thought marked a departure from the idea of group selection towards the modern notion of the gene as the unit of selection. In the mid-1970s, Motoo Kimura formulated the neutral theory of molecular evolution, firmly establishing the importance of genetic drift as a major mechanism involved in evolution.

Disciplines in Evolutionary Studies

Scholars in a number of academic disciplines and subdisciplines are involved in evolutionary studies.

Physical Anthropology

Physical anthropology emerged in the late 1800s as the study of human osteology, and the fossilized skeletal remains of other hominids. At that time, anthropologists debated whether their evidence supported Darwin's claims, because skeletal remains revealed temporal and spatial variation among hominids, but Darwin had not offered an explanation of the mechanisms that produce variation. With the recognition of Mendelian genetics and the rise of the modern synthesis, however, evolution became both the fundamental conceptual framework for, and object of study of, physical anthropologists. In addition to studying skeletal remains, they began to study genetic variation among human populations (i.e. population genetics; thus, some physical anthropologists began calling themselves biological anthropologists).

Evolutionary Biology

Evolutionary biology is a subfield of biology concerned with the origin and descent of species, as well as their change over time.

At first, it was an interdisciplinary field, including scientists from many traditional taxonomically oriented disciplines, but not a discipline in its own right. Scientists were involved who generally had specialist training in particular organisms or groups of organisms, such as mammalogy, ornithology, or herpetology, but used those organisms as systems to answer general questions in evolution. Evolutionary biology as an academic discipline in its own right emerged as a result of the modern evolutionary synthesis in the 1930s and 1940s. It was not until the 1970s and 1980s, however, that a significant number of universities had departments that specifically included the term evolutionary biology in their titles.

Evolutionary Developmental Biology

Evolutionary developmental biology is an emergent subfield of evolutionary biology that looks at genes of related and unrelated organisms. By comparing the nucleotide sequences of DNA/RNA, it is possible to experimentally develop proposals for timelines of species development. For example, gene sequences support the perspective that chimpanzees are the closest primate ancestor to humans, and that arthropods (e.g., insects) and vertebrates have a common biological ancestor.

Convergent Evolution

Convergent evolution refers to the kind of evolution wherein organisms evolve (analogous) structures or functions in spite of their evolutionary ancestors being very dissimilar or unrelated. Analogous structures pertain to those structures of unrelated (different) organisms having the same function but differing anatomical features. Because analogous structures differ in anatomy as well as developmental origin they do not implicate a common ancestral origin. The features or traits common in them evolve independently. Examples are as follows:

- The wings of bats, birds, and insects evolved independently from each other but all are used to perform the function of flying.

- The complex eyes of vertebrates, cephalopods (squid and octopus), cubozoan jellyfish, and arthropods (insects, spiders, crustaceans) evolved separately, but all perform the function of vision.

- The smelling organs of the terrestrial coconut crab are similar to those of insects.

- The very similar shells of brachiopods and bivalve molluscs.

- Prickles, thorns and spines have evolved independently to prevent or limit herbivory.

- Plant hormones such as gibberellin and abscisic acid of plants and fungi.

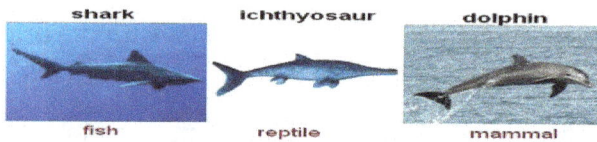

Convergent evolution

Divergent Evolution

Divergent evolution may occur as a response to changes in abiotic factors, such as a change in environmental conditions, or when a new niche becomes available. Alternatively, divergent evolution may take place in response to changes in biotic factors, such as increased or decreased pressure from competition or predation.

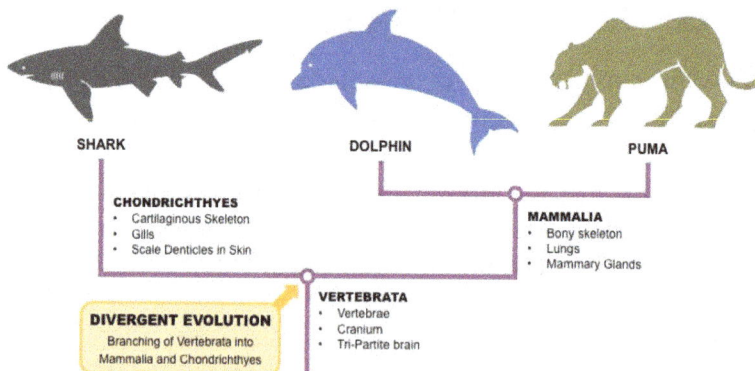

Divergent evolution is the process whereby groups from the same common ancestor evolve and accumulate differences, resulting in the formation of new species.

As selective pressures are placed upon organisms, they must develop adaptive traits in order to survive and maintain their reproductive fitness. Differences may be minor, such as the change in shape, size or function of only one structure, or they may be more pronounced and numerous, resulting in a completely different body structure or phenotype.

Divergent evolution leads to speciation, and works on the basis that there is variation within the gene pool of a population. If a reproductive barrier separates two groups within a population, different genes controlling for various aspects of an organism's ability to survive and reproduce increase or decrease in frequency as gene flow is restricted. Allopatric speciation and peripatric speciation occur when the reproductive barrier is caused by a physical or geographical barrier, such as a river or mountain range. Alternatively, sympatric speciation and parapatric speciation take place within the same geographical area.

Through divergent evolution, organisms may develop homologous structures. These are anatomically similar structures, which are present in the common ancestor and persist within the diverged organisms, although have evolved dissimilar functions.

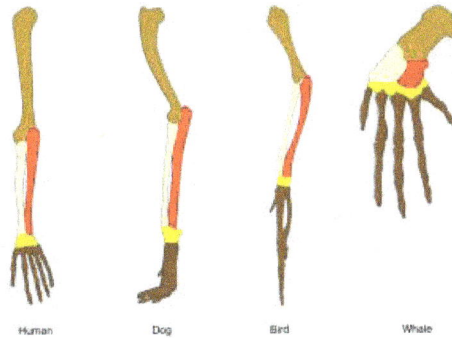

Human   Dog   Bird   Whale

The image shows an example of the homologous bones found
in the forelimb of four different types of mammal.

Examples of Divergent Evolution

Darwin's Finches

One of the most famous examples of divergent evolution was observed by Charles Darwin. Upon visiting the Galapagos Islands, Darwin noted that each of the islands had a resident population of finches belonging to the same taxonomic family. However, the bird populations on each island differed from those on nearby islands in the shape and size of their beaks.

Darwin suggested that each of the bird species had originally belonged to a single common ancestor species, which had undergone modifications of its features based on the type of food source available on each island. For example, the birds that fed on seeds and nuts evolved large crushing beaks, while cactus eaters developed longer beaks, and finer beaks evolved in birds that fed by picking insects out of trees.

When the ancestral form of finches initially colonized each island, each group contained individuals who were able to better adapt to the conditions and the available food source. These individuals survived and reproduced in their new habitat. In doing so, the genes that controlled for certain favorable aspects (e.g., longer beaks suitable for accessing nectar deep inside flowers) were spread throughout the gene pool, while the individuals without favored features died out. This is the process of natural selection.

The case of 'Darwin's Finches' (the birds actually belong to the tanager family and are not true finches) is an example of adaptive radiation, which is a form of divergent evolution.

1. Geospiza magnirostris.   2. Geospiza fortis.
3. Geospiza parvula.   4. Certhidea olivacea.

Darwin's finches

Adaptive radiation is a common feature in archipelagos such as the Galapagos Islands and Hawaii, as well as on metaphorical 'island habitats' such as mountain ranges. This is because gene flow between islands is limited when migration is not constant; however, the scale of the effect depends on the dispersal ability of the organism.

The Evolution of Primates

All of the primates on Earth evolved from a single common ancestor, most likely a primate-like, insectivore mammal, which lived around 65 million years ago in the Mesozoic Era. At that time, the world's continents were mostly connected. Fossil evidence suggests that these primitive animals lived an arboreal life, with good eyesight and hands and feet adapted to climbing through trees.

Around 55 million years ago, the first true primates evolved, diverging into the prosimians and simians.

Ancestral prosimians mostly resembled modern prosimians, which include the lemurs (endemic to Madagascar), lorises, tarsiers and bush babies. These are small-brained and relatively small-bodied, with a wet nose similar to that of a dog. They are often nocturnal, with body features that are considered 'primitive', compared to other primates.

The next big divergence occurred around 35 million years ago in the other phylogenetic branch of primates, the simians. This event resulted in the divergence of the common ancestor of all New World monkeys and Old World monkeys.

It is speculated that the two groups underwent divergent evolution as a consequence of allopatric speciation. As the continents of America and Eurasia had by this point separated, the split could have been caused by a chance migration across the Atlantic Ocean.

The New World monkeys or Platyrrhines, are native to Central and South America, as well as Mexico. They evolved flat noses and prehensile tails, which act as a fifth limb and have the ability to grasp on to trees and branches. These include familiar families such as capuchins and spider monkeys (family: Cebidae), marmosets (Callitrichidae), and howler monkeys (Atelidae).

The common ancestor of the Old World monkeys and apes split around 25 million years ago. Old World monkeys, or Catarrhini, are native to Africa and Asia, displaying a range of different adaptions to many types of habitat, from rainforests to savannah, mountains and shrubs. There are both terrestrial and arboreal Catarrhini, many of which are familiar, such as macaques genus: Macaca), baboons (Papio) and langurs (Semnopithecus).

The apes, or Hominoidea, further diverged into two groups: the lesser apes, such as gibbons (family: Hylobatidae), which are all native to Asia, and the great apes (Hominidae), which are native to Europe, Africa and Asia, and include orangutans (Genus: Pongo), gorillas (Gorilla), chimpanzees (Pan) and humans (Homo).

It is important to remember that the modern primates we see today are not evolved from each other despite their similarities (for example, great apes are not evolved from lesser apes), but that they are descended from a single common ancestor that formed two different species through divergent evolution.

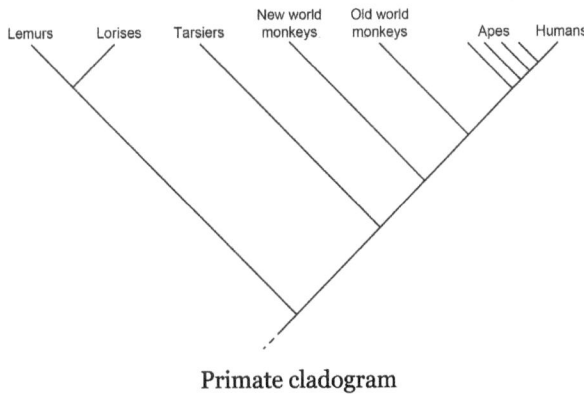

Primate cladogram

The Kit Fox and the Arctic Fox

Two species that are very closely related and have undergone divergent evolution are the kit fox (Vulpes macrotis) and the Arctic fox (Vulpes lagopus).

The kit fox is native to Western North America, and is adapted to desert environments; it has sandy coloration, and large ears, which help it to remove excess body heat.

The Arctic fox is native to Arctic regions and lives in the Arctic tundra biome of the Northern Hemisphere. Best adapted to cold climates, it has thick fur, which is white in the winter and brown in the summer, and a small, round body shape that minimizes heat loss.

Having diverged from a recent common ancestor, both these species have had to adapt to their extremely different habitats. They have evolved into two species that are clearly very distinct in terms of their ears and coats, although they still retain the majority of their ancestral features.

Parallel Evolution

Parallel evolution refers to the evolutionary process wherein two or more species in the same environment develop similar adaptation or characteristics. Example of parallel evolution: North American cactus and the African euphorbia that developed similar adaptation, which is their thick stems and sharp quills to survive the hot, arid climates. These two plant species are of different plant families but live in the same type of environment. Another example is the evolution of adaptive features between two groups of organisms living in similar habitats such as marsupial mammals in Australia and placental mammals on another continent.

Molecular Evolution

Molecular evolution is the process of the genetic material in populations of organisms changing over time. The genetic material consists of DNA, long sequences of nucleotides in each individual organism. Because most heritable changes in visible traits are a result of changes in the DNA, molecular evolution must be seen as part of general evolution. The boundary between molecular and other aspects of evolution is not clearly defined. One inequivalence is that molecular evolution takes place also in DNA with no known function (so-called "junk DNA"). Therefore the DNA of a population may "evolve molecularly", while the phenotype of descendants remains constant.

Exceptions to the General Description

Genomic imprinting (which is "epigenetic") constitutes heritability that is not coded in DNA. Evolution is prevalent also in viruses, although these are not considered to be organisms. The genetic material in viruses may consist of DNA or RNA.

Principles of Molecular Evolution

Mutations

Mutations are permanent, transmissible changes to the genetic material (usually DNA or RNA) of a cell. Mutations can be caused by copying errors in the genetic material during cell division and by exposure to radiation, chemicals, or viruses, or can occur deliberately under cellular control during the processes such as meiosis or hypermutation. Mutations are considered the driving force of evolution, where less favorable (or deleterious) mutations are removed from the gene pool by natural selection, while more favorable (or beneficial) ones tend to accumulate. Neutral mutations do not affect the organism's chances of survival in its natural environment and can accumulate over time, which might result in what is known as punctuated equilibrium; the modern interpretation of classic evolutionary theory.

Causes of Change in Allele Frequency

There are four known processes that affect the survival of a characteristic; or, more specifically, the frequency of an allele (variant of a gene):

- Mutation.

- Genetic drift describes changes in gene frequency that cannot be ascribed to selective pressures, but are due instead to events that are unrelated to inherited traits. This is especially important in small mating populations, which simply cannot have enough offspring to maintain the same gene distribution as the parental generation.

- Gene flow or gene admixture is the only one of the agents that makes populations closer genetically while building larger gene pools.

- Selection, in particular natural selection produced by differential mortality and fertility. Differential mortality is the survival rate of individuals before their reproductive age. If they survive, they are then selected further by differential fertility – that is, their total genetic contribution to the next generation. In this way, the alleles that these surviving individuals contribute to the gene pool will increase the frequency of those alleles.

The production and redistribution of variation is produced by three of the four agents of evolution: mutation, genetic drift, and gene flow.

Genome Architecture

Genome Size

Genome size is influenced by the amount of repetitive DNA as well as number of genes in an organism. The C-value paradox refers to the lack of correlation between organism 'complexity' and genome size. Explanations for the so-called paradox are two-fold. First, repetitive genetic elements can comprise large portions of the genome for many organisms, thereby inflating DNA content of the haploid genome. Secondly, the number of genes is not necessarily indicative of the number of developmental stages or tissue types in an organism. An organism with few developmental stages or tissue types may have large numbers of genes that influence non-developmental phenotypes, inflating gene content relative to developmental gene families.

Neutral explanations for genome size suggest that when population sizes are small, many mutations become nearly neutral. Hence, in small populations repetitive content and other 'junk' DNA can accumulate without placing the organism at a competitive disadvantage. There is little evidence to suggest that genome size is under strong widespread selection in multicellular eukaryotes. Genome size, independent of gene content, correlates poorly with most physiological traits and many eukaryotes, including mammals, harbor very large amounts of repetitive DNA.

However, birds likely have experienced strong selection for reduced genome size, in response to changing energetic needs for flight. Birds, unlike humans, produce nucleated red blood cells, and larger nuclei lead to lower levels of oxygen transport. Bird metabolism is far higher than that of mammals, due largely to flight, and oxygen needs are high. Hence, most birds have small, compact genomes with few repetitive elements. Indirect evidence suggests that non-avian theropod dinosaur ancestors of modern birds also had reduced genome sizes, consistent with endothermy and high energetic needs for running speed. Many bacteria have also experienced selection for small genome size, as time of replication and energy consumption are so tightly correlated with fitness.

Repetitive Elements

Transposable elements are self-replicating, selfish genetic elements which are capable of proliferating within host genomes. Many transposable elements are related to viruses, and share several proteins in common.

Chromosome Number and Organization

The number of chromosomes in an organism's genome also does not necessarily correlate with the amount of DNA in its genome. The ant *Myrmecia pilosula* has only a single pair of chromosomes whereas the Adders-tongue fern *Ophioglossum reticulatum* has up to 1260 chromosomes. Cilliate genomes house each gene in individual chromosomes, resulting in a genome which is not physically linked. Reduced linkage through creation of additional chromosomes should effectively increase the efficiency of selection.

Changes in chromosome number can play a key role in speciation, as differing chromosome numbers can serve as a barrier to reproduction in hybrids. Human chromosome 2 was created from a fusion of two chimpanzee chromosomes and still contains central telomeres as well as a vestigial second centromere. Polyploidy, especially allopolyploidy, which occurs often in plants, can also result in reproductive incompatibilities with parental species. *Agrodiatus* blue butterflies have diverse chromosome numbers ranging from n=10 to n=134 and additionally have one of the highest rates of speciation identified to date.

Gene Content and Distribution

Different organisms house different numbers of genes within their genomes as well as different patterns in the distribution of genes throughout the genome. Some organisms, such as most bacteria, *Drosophila*, and *Arabidopsis* have particularly compact genomes with little repetitive content or non-coding DNA. Other organisms, like mammals or maize, have large amounts of repetitive DNA, long introns, and substantial spacing between different genes. The content and distribution of genes within the genome can influence the rate at which certain types of mutations occur and can influence the subsequent evolution of different species. Genes with longer introns are more likely to recombine due to increased physical distance over the coding sequence. As such, long introns may facilitate ectopic recombination, and result in higher rates of new gene formation.

Organelles

In addition to the nuclear genome, endosymbiont organelles contain their own genetic material typically as circular plasmids. Mitochondrial and chloroplast DNA varies across taxa, but membrane-bound proteins, especially electron transport chain constituents are most often encoded in the organelle. Chloroplasts and mitochondria are maternally inherited in most species, as the organelles must pass through the egg. In a rare departure, some species of mussels are known to inherit mitochondria from father to son.

Origins of New Genes

New genes arise from several different genetic mechanisms including gene duplication, de novo origination, retrotransposition, chimeric gene formation, recruitment of non-coding sequence, and gene truncation.

Gene duplication initially leads to redundancy. However, duplicated gene sequences can mutate to develop new functions or specialize so that the new gene performs a subset of the original ancestral functions. In addition to duplicating whole genes, sometimes only a domain or part of a protein is duplicated so that the resulting gene is an elongated version of the parental gene.

Retrotransposition creates new genes by copying mRNA to DNA and inserting it into the genome. Retrogenes often insert into new genomic locations, and often develop new expression patterns and functions.

Chimeric genes form when duplication, deletion, or incomplete retrotransposition combine portions of two different coding sequences to produce a novel gene sequence. Chimeras often cause regulatory changes and can shuffle protein domains to produce novel adaptive functions.

De novo origin. Novel genes can also arise from previously non-coding DNA. For instance, Levine and colleagues reported the origin of five new genes in the *D. melanogaster* genome from noncoding DNA. Similar de novo origin of genes has been also shown in other organisms such as yeast, rice and humans. De novo genes may evolve from transcripts that are already expressed at low levels. Mutation of a stop codon to a regular codon or a frameshift may cause an extended protein that includes a previously non-coding sequence.

De novo evolution of genes can also be simulated in the laboratory. Donnelly et al. have shown that semi-random gene sequences can be selected for specific functions. More specifically, they selected sequences from a library that could complement a gene deletion in *E. coli*. The deleted gene encodes ferric enterobactin esterase (Fes), which releases iron from an iron chelator, enterobactin. While Fes is a 400 amino acid protein, the newly selected gene was only 100 amino acids in length and unrelated in sequence to Fes.

In vitro Molecular Evolution Experiments

Principles of molecular evolution have also been discovered, and others elucidated and tested using experimentation involving amplification, variation and selection of rapidly proliferating and genetically varying molecular species outside cells. Since the pioneering work of Sol Spiegelmann in 1967, involving RNA that replicates itself with the aid of an enzyme extracted from the Qß virus, several groups (such as Kramers and Biebricher/Luce/Eigen) studied mini and micro variants of this RNA in the 1970s and 1980s that replicate on the timescale of seconds to a minute, allowing hundreds of generations with large population sizes (e.g. 10^{14} sequences) to be followed in a single day of experimentation. The chemical kinetic elucidation of the detailed mechanism of replication meant that this type of system was the first molecular evolution system that could be fully characterised on the basis of physical chemical kinetics, later allowing the first models of the genotype to phenotype map based on sequence dependent RNA folding and refolding to be produced. Subject to maintaining the function of the multicomponent Qß enzyme, chemical conditions could be varied significantly, in order to study the influence of changing environments and selection pressures. Experiments with *in vitro* RNA quasi species included the characterisation of the error threshold for information in molecular evolution, the discovery of *de novo* evolution leading to diverse replicating RNA species and the discovery of spatial travelling waves as ideal molecular evolution reactors. Later experiments employed novel combinations of enzymes to elucidate novel aspects of interacting molecular evolution involving population dependent fitness, including work with artificially designed molecular predator prey and cooperative systems of multiple RNA and DNA. Special evolution reactors were designed for these studies, starting with serial transfer machines, flow reactors such as cellstat machines, capillary reactors, and microreactors including line flow reactors and gel slice reactors. These studies were accompanied by theoretical developments and simulations involving RNA folding and replication kinetics that elucidated the importance of the correlation structure between distance in sequence space and fitness changes, including the role of neutral networks and structural ensembles in evolutionary optimisation.

Molecular Phylogenetics

Molecular systematics is the product of the traditional fields of systematics and molecular genetics. It uses DNA, RNA, or protein sequences to resolve questions in systematics, i.e. about their correct scientific classification or taxonomy from the point of view of evolutionary biology.

Molecular systematics has been made possible by the availability of techniques for DNA sequencing, which allow the determination of the exact sequence of nucleotides or *bases* in either DNA or RNA. At present it is still a long and expensive process to sequence the entire genome of an organism, and this has been done for only a few species.

However, it is quite feasible to determine the sequence of a defined area of a particular chromosome. Typical molecular systematic analyses require the sequencing of around 1000 base pairs.

The Driving Forces of Evolution

Depending on the relative importance assigned to the various forces of evolution, three perspectives provide evolutionary explanations for molecular evolution.

Selectionist hypotheses argue that selection is the driving force of molecular evolution. While acknowledging that many mutations are neutral, selectionists attribute changes in the frequencies of neutral alleles to linkage disequilibrium with other loci that are under selection, rather than to random genetic drift. Biases in codon usage are usually explained with reference to the ability of even weak selection to shape molecular evolution.

Neutralist hypotheses emphasize the importance of mutation, purifying selection, and random genetic drift. The introduction of the neutral theory by Kimura, quickly followed by King and Jukes' own findings, led to a fierce debate about the relevance of neodarwinism at the molecular level. The Neutral theory of molecular evolution proposes that most mutations in DNA are at locations not important to function or fitness. These neutral changes drift towards fixation within a population. Positive changes will be very rare, and so will not greatly contribute to DNA polymorphisms. Deleterious mutations do not contribute much to DNA diversity because they negatively affect fitness and so are removed from the gene pool before long. This theory provides a framework for the molecular clock. The fate of neutral mutations are governed by genetic drift, and contribute to both nucleotide polymorphism and fixed differences between species.

In the strictest sense, the neutral theory is not accurate. Subtle changes in DNA very often have effects, but sometimes these effects are too small for natural selection to act on. Even synonymous mutations are not necessarily neutral because there is not a uniform amount of each codon. The nearly neutral theory expanded the neutralist perspective, suggesting that several mutations are nearly neutral, which means both random drift and natural selection is relevant to their dynamics. The main difference between the neutral theory and nearly neutral theory is that the latter focuses on weak selection, not strictly neutral.

Mutationists hypotheses emphasize random drift and biases in mutation patterns. Sueoka was the first to propose a modern mutationist view. He proposed that the variation in GC content was not the result of positive selection, but a consequence of the GC mutational pressure.

Protein Evolution

Evolution of proteins is studied by comparing the sequences and structures of proteins from many organisms representing distinct evolutionary clades. If the sequences/structures of two proteins are similar indicating that the proteins diverged from a common origin, these proteins are called as homologous proteins. More specifically, homologous proteins that exist in two distinct species are called as orthologs. Whereas, homologous proteins encoded by the genome of a single species are called paralogs.

Lipase Sequence Homology in Different Human Tissues

Query hit (click to show/hide alignment)	Target hit	Target len	Identity	Tot. score	E-value
Lipoprotein lipase (LPL) [NX_P06858-1]		475aa	100%	2570	0.0e-00
Endothelial lipase (LIPG) [NX_Q9Y5X9-1]		500aa	45%	1158	1.4e-126
Hepatic triacylglycerol lipase (LIPC) [NX_P11150-1]		499aa	43%	1037	1.5e-112
Endothelial lipase (LIPG) [NX_Q9Y5X9-2]		354aa	34%	935	1.1e-100
Pancreatic triacylglycerol lipase (PNLIP) [NX_P16233-1]		465aa	27%	503	1.2e-50
Inactive pancreatic lipase-related protein 1 (PNLIPRP1) [NX_P54315-1]		467aa	27%	497	6.4e-50
Pancreatic lipase-related protein 2 (PNLIPRP2) [NX_P54317-1]		469aa	25%	459	2.0e-46
Pancreatic lipase-related protein 3 (PNLIPRP3) [NX_Q17RR3-1]		467aa	24%	430	4.4e-42
Lipase member H (LIPH) [NX_Q8WWY8-1]		451aa	22%	423	2.9e-41
Lipase member I (LIPI) [NX_Q6XZB0-1]		460aa	21%	412	5.7e-40
Lipase member I (LIPI) [NX_Q6XZB0-2]		481aa	21%	411	6.3e-40
Lipase member I (LIPI) [NX_Q6XZB0-6]		375aa	22%	406	1.4e-39
Lipase member I (LIPI) [NX_Q6XZB0-3]		454aa	20%	406	3.1e-39

This chart compares the sequence identity of different lipase proteins throughout the human body. It demonstrates how proteins evolve, keeping some regions conserved while others change dramatically

The phylogenetic relationships of proteins are examined by multiple sequence comparisons. Phylogenetic trees of proteins can be established by the comparison of sequence identities among protoeins. Such phylogenetic trees have established that the sequence similarities among proteins reflect closely the evolutionary relationships among organisms.

Protein evolution describes the changes over time in protein shape, function, and composition. Through quantitative analysis and experimentation, scientists have strived to understand the rate and causes of protein evolution. Using the amino acid sequences of hemoglobin and cytochrome c from multiple species, scientists were able to derive estimations of protein evolution rates. What they found was that the rates were not the same among proteins. Each protein has its own rate, and that rate is constant across phylogenies (i.e., hemoglobin does not evolve at the same rate as cytochrome c, but hemoglobins from humans, mice, etc. do have comparable rates of evolution.). Not all regions within a protein mutate at the same rate; functionally important areas mutate more slowly and amino acid substitutions involving similar amino acids occurs more often than dissimilar substitutions. Overall, the level of polymorphisms in proteins seems to be fairly constant. Several species (including humans, fruit flies, and mice) have similar levels of protein polymorphism.

Relation to Nucleic Acid Evolution

Protein evolution is inescapably tied to changes and selection of DNA polymorphisms and mutations because protein sequences change in response to alterations in the DNA sequence. Amino acid sequences and nucleic acid sequences do not mutate at the same rate. Due to the degenerate nature of DNA, bases can change without affecting the amino acid sequence. For example, there are six codons that code for leucine. Thus, despite the difference in mutation rates, it is essential to incorporate nucleic acid evolution into the discussion of protein evolution. At the end of the 1960s, two groups of scientists—Kimura (1968) and King and Jukes (1969)—independently proposed that a majority of the evolutionary changes observed in proteins were neutral. Since then, the neutral theory has been expanded upon and debated.

Emergent Evolution

Emergent Evolution is an idealist concept that views development as the intermittent emergence of new and higher qualities—a process conditioned by the intervention of ideal forces. The concept came to maturity in the works of S. Alexander and of the British biologist and philosopher C. Lloyd Morgan.

Emergent evolution distinguishes between (1) quantitative changes, or "resultants," which are defined by the algebraic sum of original properties, and (2) qualitative changes, or "emergents," which cannot be reduced to the original properties and are in no way conditioned by material changes. With its gradation of "emergents," the doctrine of emergent evolution may be said to deal with "levels of existence." The number of levels of emergent evolution varies from three (matter, life, and psyche) to a score or more. The lowest level is interpreted as one that merely creates the necessary conditions for the emergence of a higher one. The nature of emergent evolution is both teleological and theological, inasmuch as certain ideal forces are said to constitute its moving force. Alexander, for example, views the moving force of emergent evolution as nisus—a striving toward something higher—and equates it with divinity as the goal of development.

Darwin's Theory of Evolution

Darwin's Theory of Evolution is the widely held notion that all life is related and has descended from a common ancestor: the birds and the bananas, the fishes and the flowers all related. Darwin's general theory presumes the development of life from non-life and stresses a purely naturalistic (undirected) "descent with modification". That is, complex creatures evolve from more simplistic ancestors naturally over time. In a nutshell, as random genetic mutations occur within an organism's genetic code, the beneficial mutations are preserved because they aid survival a process known as "natural selection." These beneficial mutations are passed on to the next generation. Over time, beneficial mutations accumulate and the result is an entirely different organism (not just a variation of the original, but an entirely different creature).

Darwin and Natural Selection

Charles Darwin

Natural selection is a process where organisms that are better adapted to an environment will survive and reproduce. This means that the advantageous alleles of this variant organism are passed on to offspring. Over many generations, the process of natural selection leads to evolution occurring.

Darwin

Charles Darwin was an English naturalist who studied variation in plants, animals and fossils during a five-year voyage around the world in the 19th century. Darwin visited four continents on the ship HMS Beagle.

Darwin observed many organisms including finches, tortoises and mocking birds, during his five week visit to the Galapágos Islands, near Ecuador in the Pacific Ocean. He continued to work and develop his ideas once he returned from his voyages. He thought that natural selection could explain how fossils related to living organisms but why there were differences between them.

Darwin proposed the theory of natural selection and how this drives the evolution of new species. Darwin is associated with the term 'survival of the fittest' which describes how natural selection works. The main stages in natural selection are:

Variation

Within a species there can be a wide range of variation, and this variation is because of differences in their genes called alleles.

Competition in a Community

All of the plants and animals in a habitat form a community. All of these living organisms need to compete with one another for survival. For example, animals will compete for food and plants compete for light.

Reproduction

Individuals with characteristics most suited to their environment are more likely to survive and reproduce - commonly known as 'survival of the fittest'. The alleles that code for a phenotype that allows these individuals to be successful within their environment are passed on to their offspring. This results in these specific alleles becoming more common.

Those that are poorly adapted to their environment are less likely to survive and reproduce. Their alleles, and therefore their phenotype, are less likely to be passed on to the next generation. Over many generations of offspring, a species will gradually evolve.

Darwin's Theory of Evolution - Slowly but Surely

Darwin's Theory of Evolution is a slow gradual process. Darwin wrote, "Natural selection acts only by taking advantage of slight successive variations; she can never take a great and sudden leap, but must advance by short and sure, though slow steps." Thus, Darwin conceded that, "If it could be demonstrated that any complex organ existed, which could not possibly have been formed by numerous, successive, slight modifications, my theory would absolutely break down." Such a complex organ would be known as an "irreducibly complex system". An irreducibly complex system is one composed of multiple parts, all of which are necessary for the system to function. If even one part is missing, the entire system will fail to function. Every individual part is integral. Thus, such a system could not have evolved slowly, piece by piece. The common mousetrap is an everyday non-biological example of irreducible complexity. It is composed of five basic parts: a catch (to hold the bait), a powerful spring, a thin rod called "the hammer," a holding bar to secure the hammer in place, and a platform to mount the trap. If any one of these parts is missing, the mechanism will not work. Each individual part is integral. The mousetrap is irreducibly complex.

Darwin's Theory of Evolution - A Theory in Crisis

Darwin's Theory of Evolution is a theory in crisis in light of the tremendous advances we've made in molecular biology, biochemistry and genetics over the past fifty years. We now know that there are in fact tens of thousands of irreducibly complex systems on the cellular level. Specified complexity pervades the microscopic biological world. Molecular biologist Michael Denton wrote, "Although the tiniest bacterial cells are incredibly small, weighing less than 10-12 grams, each is in effect a veritable micro-miniaturized factory containing thousands of exquisitely designed pieces of intricate molecular machinery, made up altogether of one hundred thousand million atoms, far more complicated than any machinery built by man and absolutely without parallel in the non-living world."

Devolution

Devolution or backward evolution is the notion that species can change into more "primitive" forms over time. In modern biology the term is redundant: evolutionary science deals with selection or adaptation that results in populations of organisms genetically different from their ancestral forms. The discipline makes no general distinction between changes leading to populations of forms less complex or more complex than their ancestors, and in such terms the concept of a primitive species cannot be defined consistently. Consequently, within the discipline such a word is rarely useful. Current non-technical application of the concept of "devolution" is based largely on the fallacies that:

- In biology there is a preferred hierarchy of structure and function, and that evolution must mean "progress" to "more advanced";

- Organisms with more complex structure and function.

Those errors in turn are related to two misconceptions: that:

- Evolution is supposed to make species more "advanced", as opposed to "primitive";

- Modern species that have lost some of the functions or complexity of their ancestors must accordingly be degenerate forms.

The concept of devolution or degenerative evolution was used by scientists in the 19th century, at this time it was believed by most biologists that evolution had some kind of direction.

In 1857 the physician Bénédict Morel influenced by Lamarckism claimed that environmental factors such as taking drugs or alcohol would produce degeneration in the offspring of those individuals, and would revert those offspring to a primitive state. Morel, a devout Catholic, had believed that mankind had started in perfection, contrasting modern humanity to the past, Morel claimed there had been "Morbid deviation from an original type". The theory of devolution, was later advocated by some biologists.

According to (Luckhurst, 2005):

> "Darwin soothed readers that evolution was progressive, and directed towards human perfectibility. The next generation of biologists were less confident or consoling. Using Darwin's theory, and many rival biological accounts of development then in circulation, scientists suspected that it was just as possible to devolve, to slip back down the evolutionary scale to prior states of development."

One of the first biologists to suggest devolution was Ray Lankester, he explored the possibility that evolution by natural selection may in some cases lead to devolution, an example he studied was the regressions in the life cycle of sea squirts. Lankester discussed the idea of devolution in his book Degeneration: A Chapter in Darwinism (1880). He was a critic of progressive evolution, pointing out that higher forms existed in the past which have since degenerated into simpler forms. Lankester argued that "if it was possible to evolve, it was also possible to devolve, and that complex organisms could devolve into simpler forms or animals".

Anton Dohrn also developed a theory of degenerative evolution based on his studies of vertebrates. According to Dohrn many chordates are degenerated because of their environmental conditions. Dohrn claimed cyclostomes such as lampreys are degenerate fish as there is no evidence their jawless state is an ancestral feature but is the product of environmental adaptation due to parasitism. According to Dohrn if cyclotomes would devolve further then they would resemble something like an Amphioxus.

Peter J. Bowler has written that devolution was taken seriously by proponents of orthogenesis and others in the late 19th century who at this period of time firmly believed that there was a direction in evolution. Orthogenesis was the belief that evolution travels in internally directed trends and levels. The paleontologist Alpheus Hyatt discussed the concept of devolution in his work, Hyatt used the concept of racial senility as the mechanism of devolution. Bowler defines racial senility as "an evolutionary retreat back to a state resembling that from which it began."

Hyatt who studied the fossils of invertebrates believed that up to a point ammonoids developed by regular stages up until a specific level but would later due to unfavourable conditions descend back to a previous level, this according to Hyatt was a form of lamarckism as the degeneration was a direct response to external factors. To Hyatt after the level of degeneration the species would then become extinct, according to Hyatt there was a "phase of youth, a phase of maturity, a phase of senility or degeneration foreshadowing the extinction of a type". To Hyatt the devolution was predetermined by internal factors which organisms can neither control or reverse. This idea of all evolutionary branches eventually running out of energy and degenerating into extinction was a pessimistic view of evolution and was unpopular amongst many scientists of the time.

Carl H. Eigenmann an ichthyologist wrote Cave vertebrates of America: a study in degenerative evolution (1909) in which he concluded that cave evolution was essentially degenerative. The entomologist William Morton Wheeler and the Lamarckian Ernest MacBride (1866-1940) also advocated degenerative evolution. According to Macbride invertebrates were actually degenerate vertebrates, his argument was based on the idea that "crawling on the seabed was inherently less stimulating than swimming in open waters."

Concepts Underlying Ideas of Devolution

The idea of de-evolution is based at least partly on the presumption that "evolution" requires some sort of purposeful direction towards "increasing complexity". Modern evolutionary theory, beginning with Darwin at least, poses no such presumption and the concept of evolutionary change is independent of either any increase in complexity of organisms sharing a gene pool, or any decrease, such as in vestigiality or in loss of genes. Earlier views that species are subject to "cultural decay", "drives to perfection", or "devolution" are practically meaningless in terms of current (neo-)Darwinian theory. Early scientific theories of transmutation of species such as Lamarckism and orthogenesis perceived species diversity as a result of a purposeful internal drive or tendency to form improved adaptations to the environment. In contrast, Darwinian evolution and its elaboration in the light of subsequent advances in biological research, have shown that adaptation through natural selection comes about when particular heritable attributes in a population happen to give a better chance of successful reproduction in the reigning environment than rival attributes do. By the same process less advantageous attributes are less "successful"; they decrease in frequency or are lost completely. Since Darwin's time it has been shown how these changes in the frequencies of attributes occur according to the mechanisms of genetics and the laws of inheritance originally investigated by Gregor Mendel. Combined with Darwin's original insights, genetic advances led to what has variously been called the modern evolutionary synthesis or neo-Darwinism. In these terms evolutionary adaptation may occur most obviously through the natural selection of particular alleles. Such alleles may be long established, or they may be new mutations. Selection also might arise from more complex epigenetic or other chromosomal changes, but the fundamental requirement is that any adaptive effect must be heritable.

The concept of devolution on the other hand, requires that there be a preferred hierarchy of structure and function, and that evolution must mean "progress" to "more advanced" organisms. For example, it could be said that "feet are better than hooves" or "lungs are better than gills", so their development is "evolutionary" whereas change to an inferior or "less advanced" structure would be called "devolution". In reality an evolutionary biologist defines all heritable changes to relative frequencies of the genes or indeed to epigenetic states in the gene pool as evolution. All gene pool changes that lead to increased fitness in terms of appropriate aspects of reproduction are seen as (neo-) Darwinian adaptation because, for the organisms possessing the changed structures, each is a useful adaptation to their circumstances. For example, hooves have advantages for running quickly on plains, which benefits horses, and feet offer advantages in climbing trees, which some ancestors of humans did.

The concept of devolution as regress from progress relates to the ancient ideas that either life came into being through special creation or that humans are the ultimate product or goal of evolution. The latter belief is related to anthropocentrism, the idea that human existence is the point of all universal existence. Such thinking can lead on to the idea that species evolve because they "need to" in order to adapt to environmental changes. Biologists refer to this misconception as teleology, the idea of intrinsic finality that things are "supposed" to be and behave a certain way, and naturally tend to act that way to pursue their own good. From a biological viewpoint, in contrast, if species evolve it is not a reaction to necessity, but rather that the population contains variations with traits that favour their natural selection. This view is supported by the fossil record which demonstrates that roughly ninety-nine percent of all species that ever lived are now extinct.

People thinking in terms of devolution commonly assume that progress is shown by increasing complexity, but biologists studying the evolution of complexity find evidence of many examples of decreasing complexity in the record of evolution. The lower jaw in fish, reptiles and mammals has seen a decrease in complexity, if measured by the number of bones. Ancestors of modern horses had several toes on each foot; modern horses have a single hooved toe. Modern humans may be evolving towards never having wisdom teeth, and already have lost most of the tail found in many other mammals - not to mention other vestigial structures, such as the vermiform appendix or the nictitating membrane. In some cases, the level of organization of living creatures can also "shift" downwards (e.g., the loss of multicellularity in some groups of protists, animals and fungi).

A more rational version of the concept of devolution, a version that does not involve concepts of "primitive" or "advanced" organisms, is based on the observation that if certain genetic changes in a particular combination (sometimes in a particular sequence as well) are precisely reversed, one should get precise reversal of the evolutionary process, yielding an atavism or "throwback", whether more or less complex than the ancestors where the process began. At a trivial level, where just one or a few mutations are involved, selection pressure in one direction can have one effect, which can be reversed by new patterns of selection when conditions change. That could be seen as reversed evolution, though the concept is of not much interest because it does not differ in any functional or effective way from any other adaptation to selection pressures. As the number of genetic changes rises however, one combinatorial effect is that it becomes vanishingly unlikely that the full course of adaptation can be reversed precisely. Also, if one of the original adaptations involved complete loss of a gene, one can neglect any probability of reversal. Accordingly, one might well expect reversal of peppered moth colour changes, but not reversal of the loss of limbs in snakes.

Dollo's Law

Complex organs evolve in a lineage over many generations, and once lost they are unlikely to re-evolve. This observation is sometimes generalized to a hypothesis known as Dollo's law, which states that evolution is not reversible. This does not mean that similar engineering solutions cannot be found by natural selection. For instance the tail of the cetacea—whales, dolphins and porpoises which are evolved from formerly land-dwelling mammals—is an adaptation of the spinal column for propulsion in water. Unlike the tail of the mammal's marine ancestor, the Sarcopterygii, and the other teleosts, which move from side to side, the cetacean's tail moves up and down as it flexes its mammalian spine, but the function of the tail in providing propulsion is remarkably similar.

Streamlining Evolution

"Devolution", the verb "devolve" and the past participle "devolved" are all common terms in science fiction for changes over time in populations of living things that make them less complex and remove some of their former adaptations. The terminology used herein is non nontechnical, but the phenomenon is a real but counterintuitive one, more accurately known as streamlining evolution. Since the development and maintenance of a feature such as an organ or a metabolite has an opportunity cost, changes in the environment that reduce the utility of an adaptation may mean that a higher evolutionary fitness is achieved by no longer using the adaptation, thus better using resources. This requires a mutation that inactivates one or more genes, perhaps by a change to DNA methylation or a methionine codon. Streamlining evolution allows evolution to remove features no longer of much/any use, like scaffolding on a completed bridge.

However, "devolution" in practice typically refers to changes that occur from a problem no longer existing rather than superior solutions existing. For instance, of the several hundred known species of animal that live their entire lives in total darkness, most have non-functional eyes rather than no eyes. This is due, for instance, to deterioration of the optic nerve. It occurs because mutations that prevent eye formation have low probability. However, several eyeless animal species, such as the Kauai cave wolf spider, who live in total darkness, and whose ancestry mostly had eyes, do exist. Together with gene duplication, streamlining evolution makes evolution surprisingly able to produce radical changes, despite being limited to successive, slight modifications.

References

- Gerardus J. H. Grubben (2004). Vegetables. PROTA. p. 404. ISBN 978-90-5782-147-9. Retrieved 10 March 2013

- Zhou Q, Zhang G, Zhang Y, et al. (2008). "On the origin of new genes in Drosophila". Genome Res. 18 (9): 1446–1455. doi:10.1101/gr.076588.108. PMC 2527705. PMID 18550802

- Evolution: newworldencyclopedia.org, Retrieved 15 March 2018

- Hanukoglu I (2017). "ASIC and ENaC type sodium channels: Conformational states and the structures of the ion selectivity filters". FEBS Journal. 284 (4): 525–545. doi:10.1111/febs.13840. PMID 27580245

- Kimura, M. (1983). The Neutral Theory of Molecular Evolution. Cambridge University Press, Cambridge. ISBN 0-521-23109-4

- Divergent-evolution: biologydictionary.net, Retrieved 10 July 2018

- Cai J, Zhao R, Jiang H, et al. (2008). "De novo origination of a new protein-coding gene inSaccharomyces cerevisiae". Genetics. 179 (1): 487–496. doi:10.1534/genetics.107.084491. PMC 2390625. PMID 18493065

- Hagen, Joel B. (1999). "Naturalists, Molecular Biologists, and the Challenge of Molecular Evolution". Journal of the History of Biology. 32 (2): 321–341. doi:10.1023/A:1004660202226. PMID 11624208

- Molecular-evolution: bio-medicine.org, Retrieved 20 June 2018

Chapter 2

Evolutionary Biology

Evolutionary biology is a branch of biology, which studies the forces of evolution that are responsible for the emergence of biological diversity on Earth. This chapter has been carefully written to provide an easy understanding of the varied facets of evolutionary biology, such as evolutionary dynamics, evolutionary developmental biology, evolution of cells and evolutionary physiology.

Evolutionary biology is a branch of biology that is primarily concerned with the evolution of species. It encompasses other fields of biology such as genetics, ecology, systematics, and paleontology. Evolution pertains to the sequence of events depicting the gradual progression of changes in the genetic composition of a biological population over successive generations. These processes resulted in the diversity of life on earth and one of the fundamental tenets is that all life on earth evolved from a common ancestor (i.e. referred to as the last universal common ancestor). At present, research in evolutionary biology includes studies in molecular evolution, genetic drift, biogeography, etc. Some of the topics currently being studied are the evolution of sexual reproduction, the evolution of cooperation, speciation, evolvability, and so on. Recently, a sub-field emerged called evolutionary developmental biology. It is a field concerned with understanding the development of an organism from a single cell to its adulthood and finding out about the phylogeny among organisms. A person studying evolutionary biology is known as an evolutionary biologist.

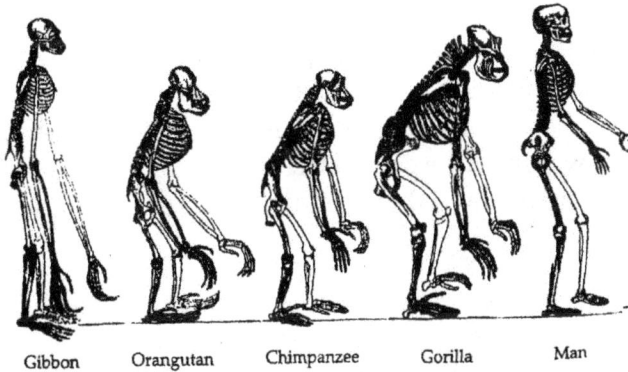

Gibbon   Orangutan   Chimpanzee   Gorilla   Man

Although gaps in the paleontological record remain, many have been filled by the researches of paleontologists since Darwin's time. Millions of fossil organisms found in well-dated rock sequences represent a succession of forms through time and manifest many evolutionary transitions. Microbial life of the simplest type (i.e., procaryotes, which are cells whose nuclear matter is not bound by a nuclear membrane) was already in existence more than three billion years ago. The oldest evidence of more complex organisms (i.e., eukaryotic cells with a true nucleus) has been discovered in flinty rocks approximately 1.4 billion years old. More advanced forms like algae, fungi, higher plants, and animals have been found only in younger geological strata. The following list presents the order in which increasingly complex forms of life appeared.

The sequence of observed forms and the fact that all (except the procaryotes) are constructed from the same basic cellular type strongly imply that all these major categories of life (including plants, algae, and fungi) have a common ancestry in the first eukaryotic cell. Moreover, there have been so many discoveries of intermediate forms between fish and amphibians, between amphibians and reptiles, between reptiles and mammals that it is often difficult to identify categorically along the line when the transition occurs from one to another particular genus or from one to another particular species. Nearly all fossils can be regarded as intermediates in some sense; they are life forms that come between ancestral forms that preceded them and those that followed.

Inferences about common descent derived from paleontology have been reinforced by comparative anatomy. The skeletons of humans, dogs, whales, and bats are strikingly similar, despite the different ways of life led by these animals and the diversity of environments in which they have flourished. The correspondence, bone by bone, can be observed in every part of the body, including the limbs: Yet a person writes, a dog runs, a whale swims, and a bat flies with structures built of the same bones. Such structures, called homologous, are best explained by common descent. Comparative anatomists investigate such homologies, not only in bone structure but also in other parts of the body as well, working out relationships from degrees of similarity.

The mammalian ear and jaw offer an example in which paleontology and comparative anatomy combine to show common ancestry through transitional stages. The lower jaws of mammals contain only one bone, whereas those of reptiles have several. The other bones in the reptile jaw are homologous with bones now found in the mammalian ear. What function could these bones have had during intermediate stages? Paleontologists have discovered intermediate forms of mammal-like reptiles (Therapsida) with a double jaw joint—one composed of the bones that persist in mammalian jaws, the other consisting of bones that eventually became the hammer and anvil of the mammalian ear.

Biogeography also has contributed evidence for common descent. The diversity of life is stupendous. Approximately 250,000 species of living plants, 100,000 species of fungi, and 1.5 million species of animals and microorganisms have been described and named, and the census is far from complete. Some species, such as human beings and our companion the dog, can live under a wide range of environmental conditions. Others are amazingly specialized. One species of the fungus Laboulbenia grows exclusively on the rear portion of the covering wings of a single species of beetle (Aphaenops cronei) found only in some caves of southern France. The larvae of the fly Drosophila carcinophila can develop only in specialized grooves beneath the flaps of the third pair of oral appendages of the land crab Gecarcinus ruricola, which is found only on certain Caribbean islands.

How can one make intelligible the colossal diversity of living beings and the existence of such extraordinary, seemingly whimsical creatures as Laboulbenia, Drosophila carcinophila, and others? Why are island groups like the Galápagos inhabited by forms similar to those on the nearest mainland but belonging to different species? Why is the indigenous life so different on different continents? The explanation is that biological diversity results from an evolutionary process whereby the descendants of local or migrant predecessors became adapted to diverse environments. For example, approximately two thousand species of flies belonging to the genus Drosophila are now found throughout the world. About one-quarter of them live only in Hawaii. More than a thousand species of snails and other land mollusks are also only found in Hawaii. The explanation for the occurrence of such great diversity among closely similar forms is that the differences resulted from adaptive colonization of isolated environments by animals with a common ancestry. The Hawaiian Islands are far from, and were never attached to, any mainland or other islands, and thus they have had few colonizers. No mammals other than one bat species lived on the Hawaiian Islands when the first human settlers arrived; very many other kinds of plants and animals were also absent. The explanation is that these kinds of organisms never reached the islands because of their great geographic isolation, while those that reached there multiplied in kind, because of the absence of related organisms that would compete for resources.

Embryology, the study of biological development from the time of conception, is another source of independent evidence for common descent. Barnacles, for instance, are sedentary crustaceans with little apparent similarity to such other crustaceans as lobsters, shrimps, or copepods. Yet barnacles pass through a free-swimming larval stage, in which they look unmistakably like other crustacean larvae. The similarity of larval stages supports the conclusion that all crustaceans have homologous parts and a common ancestry. Human and other mammalian embryos pass through a stage during which they have unmistakable but useless grooves similar to gill slits found in fishes—evidence that they and the other vertebrates shared remote ancestors that respired with the aid of gills.

The substantiation of common descent that emerges from all the foregoing lines of evidence is being validated and reinforced by the discoveries of modern biochemistry and molecular biology, a biological discipline that has emerged in the mid twentieth century. This new discipline has unveiled the nature of hereditary material and the workings of organisms at the level of enzymes and other molecules. Molecular biology provides very detailed and convincing evidence for biological evolution.

The Genetic Basis of Evolution

The central argument of Darwin's theory of evolution starts from the existence of hereditary variation. Experience with animal and plant breeding demonstrates that variations can be developed that are "useful to man." So, reasoned Darwin, variations must occur in nature that are favorable or useful in some way to the organism itself in the struggle for existence. Favorable variations are ones that increase chances for survival and procreation. Those advantageous variations are preserved and multiplied from generation to generation at the expense of less advantageous ones. This is the process known as natural selection. The outcome of the process is an organism that is well adapted to its environment, and evolution occurs as a consequence.

Biological evolution is the process of change and diversification of organisms over time, and it affects all aspects of their lives—morphology, physiology, behaviour, and ecology. Underlying these changes are changes in the hereditary material (DNA). Hence, in genetic terms, evolution consists of changes in the organism's hereditary makeup. Natural selection, then, can be defined as the differential reproduction of alternative hereditary variants, determined by the fact that some variants increase the likelihood that the organisms having them will survive and reproduce more successfully than will organisms carrying alternative variants. Selection may be due to differences in survival, in fertility, in rate of development, in mating success, or in any other aspect of the life cycle. All these differences can be incorporated under the term differential reproduction because all result in natural selection to the extent that they affect the number of progeny an organism leaves.

Evolution can be seen as a two-step process. First, hereditary variation takes place; second, selection occurs of those genetic variants that are passed on most effectively to the following generations. Hereditary variation also entails two mechanisms: the spontaneous mutation of one variant to another, and the sexual process that recombines those variants to form a multitude of variations.

The information encoded in the nucleotide sequence of DNA is, as a rule, faithfully reproduced during replication, so that each replication results in two DNA molecules that are identical to each other and to the parent molecule. But occasionally "mistakes," or mutations, occur in the DNA molecule during replication, so that daughter molecules differ from the parent molecules in at least one of the letters in the DNA sequence. Mutations can be classified into two categories: gene, or point, mutations, which affect one or only a few letters (nucleotides) within a gene; and chromosomal mutations, which either change the number of chromosomes or change the number or arrangement of genes on a chromosome. Chromosomes are the elongated structures that store the DNA of each cell.

Newly arisen mutations are more likely to be harmful than beneficial to their carriers, because mutations are random events with respect to adaptation; that is, their occurrence is independent of any possible consequences. Harmful mutations are eliminated or kept in check by natural selection. Occasionally, however, a new mutation may increase the organism's adaptation. The probability of such an event's happening is greater when organisms colonize a new territory or when environmental changes confront a population with new challenges. In these cases there is greater opportunity for new mutations to be better adaptive. The consequences of mutations depend on the environment. Increased melanin pigmentation may be advantageous to inhabitants of tropical Africa, where dark skin protects them from the Sun's ultraviolet radiation; but it is not beneficial in Scandinavia, where the intensity of sunlight is low and light skin facilitates the synthesis of vitamin D.

Mutation rates are low, but new mutants appear continuously in nature because there are many individuals in every species and many genes in every individual. More important is the storage of variation, arisen by past mutations. Thus, it is not surprising to see that when new environmental challenges arise, species are able to adapt to them. More than two hundred insect species, for example, have developed resistance to the pesticide DDT in different parts of the world where spraying has been intense. Although the insects had never before encountered this synthetic compound, they adapted to it rapidly by means of mutations that allowed them to survive in its presence. Similarly, many species of moths and butterflies in industrialized regions have shown an increase in the frequency of individuals with dark wings in response to environmental pollution, an adaptation known as industrial melanism.

Dynamics of Genetic Change

The genetic variation present in natural populations of organisms is sorted out in new ways in each generation by the process of sexual reproduction. But heredity by itself does not change gene frequencies. This principle is formally stated by the Hardy-Weinberg law, an algebraic equation that describes the genetic equilibrium in a population.

The Hardy-weinberg law plays in evolutionary studies a role similar to that of Isaac Newton's first law of motion in mechanics. Newton's first law says that a body not acted upon by a net external force remains at rest or maintains a constant velocity. In fact, there are always external forces acting upon physical objects (gravity, for example), but the first law provides the starting point for the application of other laws. Similarly, organisms are subject to mutation, selection, and other processes that change gene frequencies, and the effects of these processes are calculated by using the Hardy-Weinberg law as the starting point. There are four processes of gene frequency change: mutation, migration, drift, and natural selection.

Mutations change gene frequencies very slowly, since mutation rates are low. Migration, or gene flow, takes place when individuals migrate from one population to another and interbreed with its members. The genetic make-up of populations changes locally whenever different populations intermingle. In general, the greater the difference in gene frequencies between the resident and the migrant individuals, and the larger the number of migrants, the greater effect the migrants have in changing the genetic constitution of the resident population.

Moreover, gene frequencies can change from one generation to another by a process of pure chance known as genetic drift. This occurs because populations are finite in numbers, and thus the frequency of a gene may change in the following generation by accidents of sampling, just as it is possible to get more or less than fifty "heads" in one hundred throws of a coin simply by chance. The magnitude of the gene frequency changes due to genetic drift is inversely related to the size of the population; the larger the number of reproducing individuals, the smaller the effects of genetic drift. The effects of genetic drift from one generation to the next are quite small in most natural populations, which generally consist of thousands of reproducing individuals. The effects over many generations are more important. Genetic drift can have important evolutionary consequences when a new population becomes established by only a few individuals, as in the colonization of islands and lakes. This is one reason why species in neighboring islands, such as those in the Hawaiian archipelago, are often more heterogeneous than species in comparable continental areas adjacent to one another.

Gradual and Punctuational Evolution

Morphological evolution is by and large a gradual process, as shown by the fossil record. Major evolutionary changes are usually due to a building up over the ages of relatively small changes. But the fossil record is discontinuous. Fossil strata are separated by sharp boundaries; accumulation of fossils within a geologic deposit (stratum) is fairly constant over time, but the transition from one stratum to another may involve gaps of tens of thousands of years. Different species, characterized by small but discontinuous morphological changes, typically appear at the boundaries between strata, whereas the fossils within a stratum exhibit little morphological variation. That is not to say that the transition from one stratum to another always involves sudden changes in morphology; on the contrary, fossil forms often persist virtually unchanged through several geologic strata, each representing millions of years.

According to some paleontologists the frequent discontinuities of the fossil record are not artefacts created by gaps in the record, but rather reflect the true nature of morphological evolution, which happens in sudden bursts associated with the formation of new species. This proposition is known as the punctuated equilibrium model of morphological evolution. The question whether morphological evolution in the fossil record is predominantly punctuational or gradual is a subject of active investigation and debate. The argument is not about whether only one or the other pattern ever occurs; it is about their relative frequency. Some paleontologists argue that morphological evolution is in most cases gradual and only rarely jerky, whereas others think the opposite is true. Much of the problem is that gradualness or jerkiness is in the eye of the beholder.

DNA and Protein Evolution

The advances of molecular biology have made possible the comparative study of proteins and the nucleic acid DNA, which is the repository of hereditary (evolutionary and developmental) information. Nucleic acids and proteins are linear molecules made up of sequences of units—nucleotides in the case of nucleic acids, amino acids in the case of proteins—which retain considerable amounts of evolutionary information. Comparing macromolecules from two different species establishes the number of their units that are different. Because evolution usually occurs by changing one unit at a time, the number of differences is an indication of the recency of common ancestry. Changes in evolutionary rates may create difficulties, but macromolecular studies have two notable advantages over comparative anatomy and other classical disciplines. One is that the information is more readily quantifiable. The number of units that are different is precisely established when the sequence of units is known for a given macromolecule in different organisms. The other advantage is that comparisons can be made even between very different sorts of organisms. There is very little that comparative anatomy can say when organisms as diverse as yeasts, pine trees, and human beings are compared; but there are homologous DNA and protein molecules that can be compared in all three.

Informational macromolecules provide information not only about the topology of evolutionary history, but also about the amount of genetic change that has occurred in any given branch. Studies of molecular evolution rates have led to the proposition that macromolecules evolve at fairly constant rates and, thus, that they can be used as evolutionary clocks, in order to determine the time when the various branching events occurred. The molecular evolutionary clock is not a metronomic clock, like a watch or other timepiece that measures time exactly, but a stochastic clock like radioactive decay. In a stochastic clock, the probability of a certain amount of change is constant, although some variation occurs in the actual amount of change. Over fairly long periods of time, a stochastic clock is quite accurate. The enormous potential of the molecular evolutionary clock lies in the fact that each gene or protein is a separate clock. Each clock "ticks" at a different rate—the rate of evolution characteristic of a particular gene or protein—but each of the thousands of genes or proteins provides an independent measure of the same evolutionary events.

Evolutionists have found that the amount of variation observed in the evolution of DNA and proteins is greater than is expected from a stochastic clock; in other words, the clock is inaccurate. The discrepancies in evolutionary rates along different lineages are not excessively large, however. It turns out that it is possible to time phylogenetic events with accuracy, but more genes or proteins must be examined than would be required if the clock were stochastically accurate. The average rates obtained for several DNA sequences or proteins taken together provide a fairly precise clock, particularly when many species are investigated.

Evolutionary Developmental Biology

Evolutionary developmental biology (evo–devo) is that part of biology concerned with how changes in embryonic development during single generations relate to the evolutionary changes that occur between generations. Charles Darwin argued for the importance of development (embryology) in understanding evolution. After the discovery in 1900 of Mendel's research on genetics, however, any relationship between development and evolution was either regarded as unimportant for understanding the processes of evolution or as a black box into which it was hard to see. Research over the past two decades has opened that black box, revealing how studies in evo–devo highlight the mechanisms that link genes (the genotype) with structures (the phenotype). This is vitally important because genes do not make structures. Developmental processes make structures using road maps provided by genes, but using many other signals as well—physical forces such as mechanical stimulation, temperature of the environment, and interaction with chemical products produced by other species—often species in entirely different kingdoms as in interactions between bacteria and squid or between leaves and larvae. Not only do genes not make structures (the phenotype), but new properties and mechanisms emerge during embryonic development: genes are regulated differentially in different cells and places; aggregations of similar cells provide the cellular resources (modules) from which tissues and organs arise; modules and populations of differently differentiated cells interact to set development along particular tracks; and organisms interact with their environment and create their niche in that environment. Such interactions are often termed "epigenetic," meaning that they direct gene activity using mechanisms that are not encoded in the DNA of the genes.

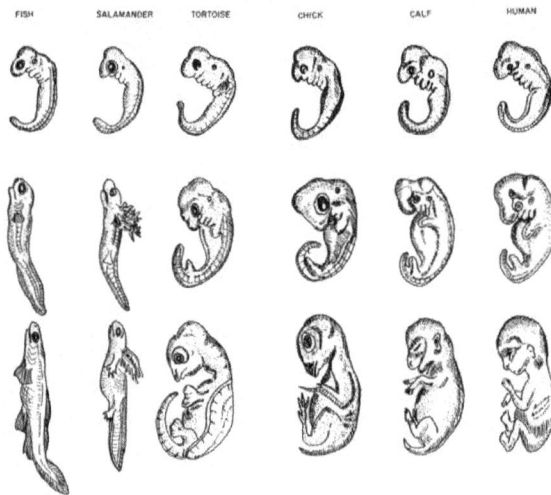

Origins of Evo-Devo

Embryos provided the way to study evolution. The fossil record was incomplete. Embryos, on the other hand, recorded in their development the history of their ancestors. This history had to be read with great care; there were gaps in the record, and secondary specializations such as the placenta could confuse the unwary. Nevertheless, from the late 1860s or early 1870s until the mid-1880s, evolutionary embryology was the field that attracted the brightest and best zoologists.

Bateson commented that "Morphology was studied because it was the material believed to be the most favorable for the elucidation of the problems of evolution, and we all thought that in embryology the quintessence of morphological truth was most palpably presented. Therefore every aspiring zoologist was an embryologist, and the one topic of professional conversation was evolution".

Frustration with reconstructing evolutionary trees from embryonic sequences, the rise of experimental and physiological approaches to embryonic development in the 1880s, and the rediscovery of Mendelian genetics in 1900 all cast evolutionary embryology into a backwater from which it would take a century to resurface. Mendel's principles of segregation and assortment coupled with studies on the fruit fly Drosophila provided a powerful foundation upon which the new science of genetics was built.

By the middle of the twentieth century, maintenance of the features of organisms, variation in those features, and the origin of new features all seemed explicable by a fusion of Mendelian and population genetics. Paradoxically, the use of Drosophila as the model organism for genetics eliminated the roles of embryonic development and of the environment from evolutionary discussion and theory; inbred laboratory organisms display none of the variation and adaptability seen in nature.

Evo–devo exploded as heterochrony was found everywhere. Along the way, heterochrony became such a pervasive term that it lost some of its explanatory power; any change in timing became heterochrony, whether evolutionarily relevant or not.

The next major impetus to evo–devo was not the resurrection of a previously known evolutionary developmental mechanisms but the discovery that all animals (subsequently shown for all plants and fungi too) share genes that contain a 180-bp sequence known as the homeobox and that these genes, known as homeobox, homeotic, or Hox genes, are responsible for determining that animals have an anterior and a posterior, a dorsal and a ventral side, and specific regions (often as repeated segments) along the body axis—head at one end tail at the other, thorax in front of abdomen, wings on a specific pair of segments, and so forth.

"Master" genes, also known as developmental or regulatory genes, were discovered. One of the best understood of such genes is the paired-box protein gene known as Pax-6 in vertebrates and as eyeless (ey) in Drosophila. As determined from its DNA sequence, orthologues of Pax-6 are present throughout the Animal Kingdom. As determined from functional studies, the role of Pax-6 in anteriorizing the embryo and in the formation of anterior sensory structures also is conserved across the Animal Kingdom. Although best known as the major gene controlling eye development, Pax-6 functions in organisms that lack eyes, reflecting its ancient developmental role. Here is an astonishing and previously unthought-of genetic conservation across animals whose morphology varied enormously.

Heterochrony, homeotic genes, increasingly resolved relationships between organisms (phylogenetic trees), and an appreciation during the 1980s and 1990s of the importance of ecological and species interactions led evo–devo to the position where the aims of evo–devo could be stated as understanding:

- The origination and evolution of embryonic development;

- The role of modifications of developmental processes in the production of novel features;

- How the adaptive plasticity of development facilitates the origin and maintenance of complex life cycles with embryos, larvae, and adults;

- How developing organisms interact with their ecological environment to facilitate evolutionary change.

Others had similar lists. Müller (2007a, b) and Collins et al. (2007) listed seven approaches and aims of evo–devo:

- The origin of developmental systems;

- The evolution of developmental systems;

- Modifications of timing and context of developmental processes;

- Environment–development interactions;

- Maintenance of phenotypic variation;

- Origin of phenotypic novelty;

- Integration of genetics and epigenetic mechanisms.

Molecular genetics has revolutionized evo–devo over the past two decades. Integrating our expanding molecular understanding with mechanisms operating at the cell or other levels (tissues, organs, whole organism, organism–environment interactions; Box 1) has been and remains a major goal and a challenge for evo–devo. Essentially, it involves opening the black box between genotype and phenotype, taking out what is found in the box and how it fits together and then determining how to put the contents back in the box. One of the items found in the black box is known as epigenetics.

A sample of evolutionary developmental mechanisms operating at various levels:

Gene	Regulation, networks, interactions, genome size, epigenetic processes (methylation, imprinting, chromosome inactivation)
Cell	Division, migration, condensation, differentiation, interaction, patterning, morphogenesis, embryonic induction

Tissue, organ	Differentiation, specialization, embryonic inductions, epithelial–mesenchymal interactions, growth
Organism	Ontogenetic re-patterning, genetic assimilation, phenotypic plasticity, polymorphism, functional morphology
Environment	Phenotypic responses to chemicals released by predators, prey, and food supplies; temperature; crowding

Epigenetics

Many of the controls on gene regulation and function are subsumed under the term "epigenetics," a term coined by the British geneticist and embryologist Conrad Waddington for the causal factors that control gene action during development defined epigenetics as "the sum of the genetic and non-genetic factors acting upon cells to control selectively the gene expression that produces increasing phenotypic complexity during development". To this, I would only add "and evolution" at the end.

Epigenetics is still used with this original meaning but has increasingly come to be applied at the molecular level for heritable changes to the DNA other than changes to the nucleotide bases. Such changes include methylation, imprinting, and regulation of chromatin—which have been known for some time and regulation of genes by small RNA molecules, especially miRNAs. This now very well-characterized second inheritance system was appreciated in its infancy by John Maynard Smith, a leading twentieth century evolutionary theorist: "There is a second inheritance system—an epigenetic inheritance system—in addition to the system based on DNA sequence that links sexual generations.

The heritable aspect of epigenetics has shown us that organisms do not start their lives as naked nuclear DNA. They possess DNA in their mitochondria, epigenetic "marks" in their nuclear DNA, and they inherit mRNA and proteins that were produced under the control of their mother's DNA and deposited into the egg cytoplasm. Epigenetics provides another, but not the only other, means by which heritable information operates in organismal development.

Integrated Mechanisms

Integrated studies using molecular biology, molecular genetics, developmental biology, phylogenetics, paleontology, and molecular paleobiology are revealing previously unimagined information on how features change during evolution.

One instance is exemplified by the rise of paleobiology as a discipline that brought evolutionary theory back into paleontology and incorporated developmental, phylogenetic, and environmental approaches into a biological perspective of fossils.

New groups of organisms, as represented by new classes or orders, can arise slowly through the gradual accumulation of new characters or can arise more rapidly through key innovations (the origination of wings or lungs, for example) or through coordinated changes in different characters as seen in the origin of lungs, middle ear ossicles, and the transformation of fins to limbs at the origin of the tetrapods. This is how microevolutionary changes within species are linked to macroevolutionary changes, as reflected in levels of classification: "the careful analyses of the differences in pathways between organisms of known phylogenetic relationship".

The list of mechanisms summarized in Box implies that evolutionary developmental mechanisms are not all found in the genes, although all have a genetic basis. This is because new mechanisms emerge as development proceeds. Evolutionary developmental mechanisms may be genetic, cellular, developmental, physiological, hormonal, or any combination of these. Embryonic development is hierarchical, with new properties and mechanisms emerging as development unfolds, each dependent on the stage/processes preceding them. The single cell that is the fertilized egg cannot show any of the cell-to-cell interactions that characterize the multicellular embryo, some of which come about because cells take up new positions in the embryo through active migration. The recent elucidation of gene networks is providing the regulatory link between the genotype and cellular modules. Embryonic inductions, tissue, organ, and functional interactions link cellular activity to the phenotype. The onset of embryonic movement ushers in a new type of process, which are interactions between developing tissues and organs such as bones and muscles.

Evolutionary Dynamics

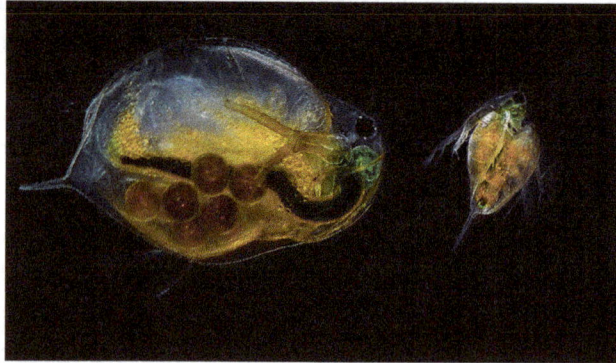

Evolutionary change is a hallmark of biological systems. We are interested in the evolutionary dynamics of pathogen populations, specifically RNA viruses, such as HIV and HCV, and tumour cell populations. These systems have in common that the evolutionary changes occur on a time scale that is observable and that the evolutionary dynamics are responsible for clinical outcomes of the disease. Using evolutionary modelling and population genetics, we have, for example, modelled the evolution of two competing strains in a virus population and the genetic progression of cancer.

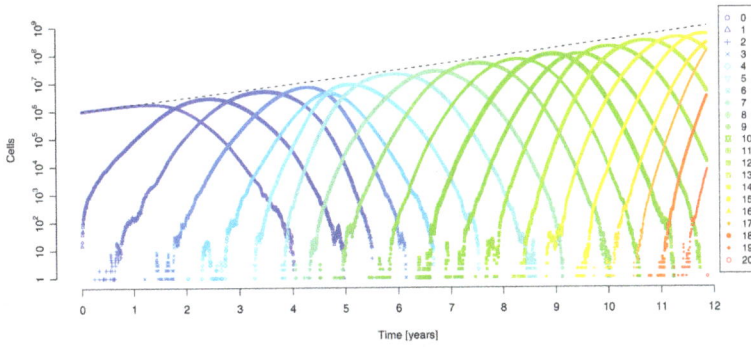

Traveling mutant waves in a simulated evolving tumour

Many disease-causing evolutionary processes of pathogenic agents occur repeatedly, for example, the same virus infecting the same type of host organism. Data from such reproducible evolutionary processes can be used to learn structural constraints on the available evolutionary pathways and on the dynamics genetic adaptation. We develop probabilistic graphical models to describe the accumulation of genetic alterations. This approach has been applied to predict the development of drug resistance in HIV and the occurrence of cancer-driving mutations in tumours.

Darwinian evolution is based on three fundamental principles, reproduction, mutation and selection which describe how populations change over time and how new forms evolve out of old ones. There are numerous mathematical descriptions of the resulting evolutionary dynamics.

The Quasispecies Equation

Let us start with the quasispecies equation of molecular evolution. The variable x_i denotes the relative abundance of a genetic sequence, i; in a population. The fitness, f_i; of this sequence is

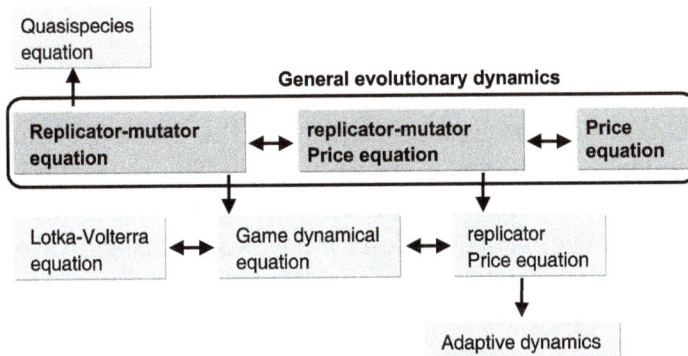

Figure: General evolutionary dynamics are described by the equivalence between the replicator–mutator equation, the replicator–mutator Price equation (which uses the labeling system of the replicator–mutator framework) and the Price equation.

The game dynamical equation is obtained from the replicator–mutator equation by

neglecting mutation. Similarly, the replicator Price equation is derived from the Price equation in the absence of mutation. The Lotka– Volterra equation, the game dynamical equation and the replicator Price equation are equivalent. Adaptive dynamics can be derived from the replicator Price equation. The quasispecies equation is a special case of the replicator–mutator equation for the case of constant fitness.

Determined by its replication rate, the average fitness of the population is given by $\bar{f} = \sum_i f_i x_i$ There are n genetic sequences. Replication is error-prone; the probability that replication of sequence i gives rise to sequence j is given by q_{ij}: These quantities describe the mutation matrix, Q: The rate at which xi changes over time is given by,

$$\dot{x}_i = \sum_{j=1}^{n} x_j f_j q_{ji} - x_i \bar{f}.$$

The term, $-x_i \bar{f}$, ensures that $\sum_i x_i = 1$. The quasispecies equation describes the adaptation of a population on a constant fitness landscape. The underlying geometry is given by sequence space, which is a high-dimensional array of sequences arranged in such a way that neighboring sequences differ by a single-point mutation.

Evolutionary Game Dynamics and Lotka–Volterra

Evolutionary game theory is a phenotypic approach to evolutionary dynamics. It describes the natural selection of strategies in evolutionary games, such as the Hawk–Dove game or the Prisoner's Dilemma.

The key aspect of evolutionary game theory is frequency-dependent selection: the fitness of an individual depends on the frequency of other strategies in the population, Let x_i denote the frequency of strategy i; Its fitness, $f_i(x)$; is a function of the distribution of the population given by the vector $x = (x_1, ..., x_n)$. Evolutionary game dynamics of discrete phenotypes are described by the replicator equation.

$$\dot{x}_i = x_i \left[f_i(x) - \bar{f} \right]$$

In ecology, the Lotka–Volterra equation describes the interaction among n different species. The abundance of species i is given by y_i: The reproductive rate (fitness), f_i; of each species depends on the abundance of other species. In general, we have $\dot{y}_i = y_i f_i(y)$: The Lotka–Volterra equation for n - 1 species is equivalent to the replication equation for n phenotypes. Let $y = \sum_{i=1}^{n-1} y_i$: The equivalence can be shown with the transformation $x_i = y_i / (1+y)$ for $i = 1, ..., n-1$ and $x_n = 1/(1+y)$.

The Replicator–Mutator Equation

The quasispecies equation lacks frequency dependent selection, while the replicator equation lacks mutation. Combining these two equations we obtain.

$$\dot{x}_i = \sum_{j=1}^{n} x_j f_j(x) q_{ji} - x_i \bar{f}$$

This "replicator–mutator equation" describes both frequency-dependent selection and mutation. It has been used in population genetics, autocatalytic reaction networks, game theory and language evolution. It is clear that the quasispecies equation and the replicator equation are special cases of the replicator–mutator equation.

The Price Equation

In 1970, George Price derived an equation to describe any form of selection. The "Price equation" was used by Hamilton in his seminal work on kin selection. It has been applied to problems in evolutionary genetics, social evolution, group selection, sex ratio and ecological diversity. Fisher's fundamental theorem of natural selection can be directly derived from the Price equation. For continuous time, the Price equation is of the form.

$$\dot{E}(P) = Cov(f,p) + E(\dot{p})$$

The numerical value of an arbitrary trait of individual i is given by p_i. The population P average of this trait is given by $\bar{p} \equiv E(p) = \sum_i p_i x_i$. The covariance of trait p and fitness f is given by $Cov(f,p) = \sum_i x_i f_i p_i - \bar{f}\bar{p}$. The second term equation $\dot{E}(P) = Cov(f,p) + E(\dot{p})$ is the population average of the rate at which the trait values change over time. We have $E(\dot{p}) = \sum_i x_i \dot{P}_i$. If the trait values, p_i, do not change with time, we obtain the "covariance equation", $\dot{E}(p) = Cov(f,p)$.

Equivalence

We will now show that the game-dynamical equation is equivalent to the Price equation, while the replicator–mutator equation is equivalent to an expanded Price equation of the form,

$$\dot{E}(p) = Cov(f,p) + E(\dot{p}) + E(f \Delta_m P)$$

The additional term $E(f\Delta_m P) = \sum_i x_i f_i \Delta_m P_i$ describes mutation among types, with $\Delta_m P_i = \sum_j q_{ij}(P_j - P_i)$ denoting the expected change in trait value when mutating from type i.

We have $\bar{p} \equiv E(p) = \sum_i p_i x_i$ and therefore $\dot{E}(p) = \sum_i P_i \dot{x}_i + \sum_i x_i \dot{p}_i$. From the replicator–mutator equation, we obtain,

$$\dot{E}(p) = \sum_i P_i \left(\sum_j x_j f_j q_{ji} - x_i \bar{f} \right) + E(\dot{P})$$

$$= \sum_{ij} p_i x_i f_j q_{ji} - \overline{fp} + E(\dot{p})$$

$$= \sum_{ij} p_i x_i f_j q_{ji} - \overline{fp}$$

$$+ \sum_{ij} (p_i - p_j) x_j f_j q_{ji} + E(\dot{P})$$

$$= \sum_{ij} p_i x_i f_j - \overline{fp} + \sum_j x_j f_j \sum_i q_{ji}(p_i - p_j)$$

$$+ E(\dot{P}) = Cov(f,p) + E(\dot{p}) + E(f\Delta_m p)$$

This is the expanded Price equation with the additional mutation term. The term $Cov(f,p)$ describes selection (and in our framework would include epistatic and dominance interactions among genes and alleles). The term $E(\dot{p})$ describes changes in trait value, which may be consequences of changes in the environment or the trait being frequency-dependent (such as fitness). The term $E(f\Delta_m P)$ describes mutation among the different types.

In the same way, we can derive the standard Price equation from the replicator eqn. In general, the Price equation is dynamically insufficient. To calculate how the population average of a trait changes with time, we need to consider a differential equation for the covariance, which in turn will include higher moments. Dynamic sufficiency can only be obtained in special cases. For our purpose, the trick is to consider n traits that are indicator functions of the n types: for $i = 1...n$ we have $P_j^{(i)} = 1$ if i ¼ j and $P_j^{(i)} = 0$ if $i \neq j$: The population average of $P^{(i)}$ is $E(p^{(i)}) = \sum_j x_j p_j^{(i)} = x_i$.

For the three terms on the RHS of the expanded Price equation, we obtain $Cov(f, p^{(i)}) = x_i f_i - \overline{f}), E(\dot{p}^{(i)}) = 0$ and $E(f\Delta_m p^{(i)}) = -x_i f_i + \sum_k x_k f_k q_{ki}$ .Hence,

The Price equation for trait $p^{(i)}$ leads to the replicator–mutator equation for frequency x_i:

We also show the equivalence between the replicator–mutator equation and the expanded Price equation for discrete time and for sexual reproduction with recombination. One can also show equivalence between the Price equation and a replicator–mutator equation describing continuous phenotypes.

In our framework, the Price equation does not include mutation, whereas the expanded Price equation does include mutation. It is interesting to note that Price does not mention mutation in his original papers. Nevertheless, Frank points out that the Price equation is an exact and complete description of evolutionary dynamics including both selection and mutation. The resolution of this apparent discrepancy lies in the different labelling systems. In our paper, we use the labelling system of the replicator–mutator framework $x_i t$ denotes the relative abundance of type i individuals at time t: In the original Price equation, however, $x_i t$ denotes the relative abundance of individuals at time t that are derived from type I individuals at time 0. Indeed, with this unusual labeling system the Price equation can be interpreted to include any form of mutation and selection. Hence, it makes sense to call "replicator Price equation" and "replicator–mutator Price equation" if one uses the labeling system of the replicator framework. In contrast, "Price equation" should refer to labeling system of Price. In this sense, the Price equation is equivalent to the replicator–mutator Price equation which is equivalent to the replicator–mutator equation.

Adaptive Dynamics

Another framework for evolutionary change is given by adaptive dynamics, which describes how continuous traits or strategies change under mutation and frequency-dependent selection. Adaptive dynamics assume there is a resident population which is surrounded by a cloud of mutants. Selection chooses the mutant with maximum fitness in the context of the resident population. In the limit of many mutants very close to the resident population, one obtains an equation using partial derivatives of trait values. Adaptive dynamics illustrate the nature of evolutionarily stable strategies, which emerge as stable or unstable equilibrium points. There is also a connection to Wright's "adaptive landscape" and formulations of selection gradients. It turns out that adaptive dynamics can be derived from the Price equation. Let us assume that the population is described by a continuous distribution, $x(p)$; of some trait variable, p The fitness of individuals with a particular trait value p depends on x and is given by $f(p;x)$

Let us start from the expanded Price equation but assume that the expect mutational change for trait p is zero. Hence, $E(f\Delta_m P) = 0$ and we obtain the Price equation. Observe, however, that we do not assume there is no mutation, we only assume that on average mutational events are equally likely to increase or decrease the trait value. Let us further assume that the trait values, pi; do not change with time. Hence, $E(\dot{P}) = 0$ and we obtain the covariance equation, $\dot{E}(p) = Cov[p, f(p;x)]$. The fitness of individuals with a trait value p is given by $f(p;x)$ where x describes the distribution of the population. A first-order Taylor expansion of $f(p;x)$ around the population average $\bar{p} \equiv E(p) = \int x(p)p\,dp$ is given by,

$$f(p;x) \approx f(\bar{p};x) + (p - \bar{p}) \frac{\partial f(q;x)}{\partial q}\Big|q = \bar{q}$$

Hence,

$$\dot{E}(p) \approx Cov(p, f(\bar{p};x) + (p - \bar{p})\frac{\partial f(q;x)}{\partial q}\Big|_{q=\bar{p}})$$

$$= Var(p)\frac{\partial f(q:x)}{\partial q}\Big|_{q=\bar{p}}$$

This equation describes the adaptive dynamics for trait p.

Evolution of Cells

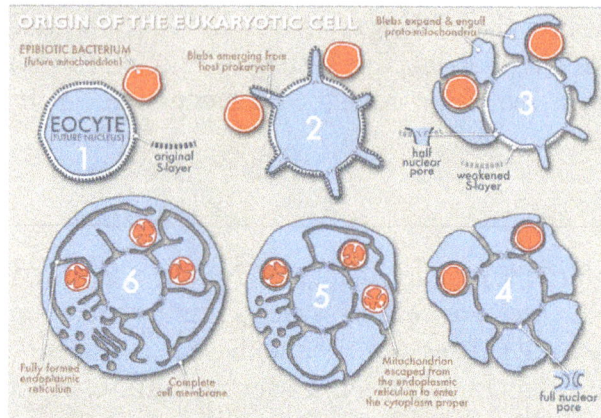

Early stages in the evolutionary pathway of cells presumably centred on RNA molecules, which not only present specific catalytic surfaces but also contain the potential for their own duplication through the formation of a complementary RNA molecule. It is assumed that a small RNA molecule eventually appeared that was able to catalyze its own duplication.

Imperfections in primitive RNA replication likely gave rise to many variant autocatalytic RNA molecules. Molecules of RNA that acquired variations that increased the speed or the fidelity of self-replication would have out multiplied other, less-competent RNA molecules. In addition, other small RNA molecules that existed in symbiosis with autocatalytic RNA molecules underwent natural selection for their ability to catalyze useful secondary reactions such as the production of better precursor molecules. In this way, sophisticated families of RNA catalysts could have evolved together, since cooperation between different molecules produced a system that was much more effective at self-replication than a collection of individual RNA catalysts.

Another major step in the evolution of the cell would have been the development, in one family of self-replicating RNA, of a primitive mechanism of protein synthesis. Protein molecules cannot provide the information for the synthesis of other protein molecules like themselves. This information must ultimately be derived from a nucleic acid sequence. Protein synthesis is much more complex than RNA synthesis, and it could not have arisen before a group of powerful RNA catalysts evolved. Each of these catalysts presumably has its counterpart among the RNA molecules that function in the current cell:

(1) There was an information RNA molecule, much like messenger RNA (mRNA), whose nucleotide sequence was read to create an amino acid sequence;

(2) There was a group of adaptor RNA molecules, much like transfer RNA (tRNA), that could bind to both mRNA and a specific activated amino acid;

(3) Finally, there was an RNA catalyst, much like ribosomal RNA (rRNA), that facilitated the joining together of the amino acids aligned on the mRNA by the adaptor RNA.

At some point in the evolution of biological catalysts, the first cell was formed. This would have required the partitioning of the primitive biological catalysts into individual units, each surrounded by a membrane. Membrane formation might have occurred quite simply, since many amphiphilic molecules—half hydrophobic (water-repelling) and half hydrophilic (water-loving)—aggregate to form bilayer sheets in which the hydrophobic portions of the molecules line up in rows to form the interior of the sheet and leave the hydrophilic portions to face the water. Such bilayer sheets can spontaneously close up to form the walls of small, spherical vesicles, as can the phospholipid bilayer membranes of present-day cells.

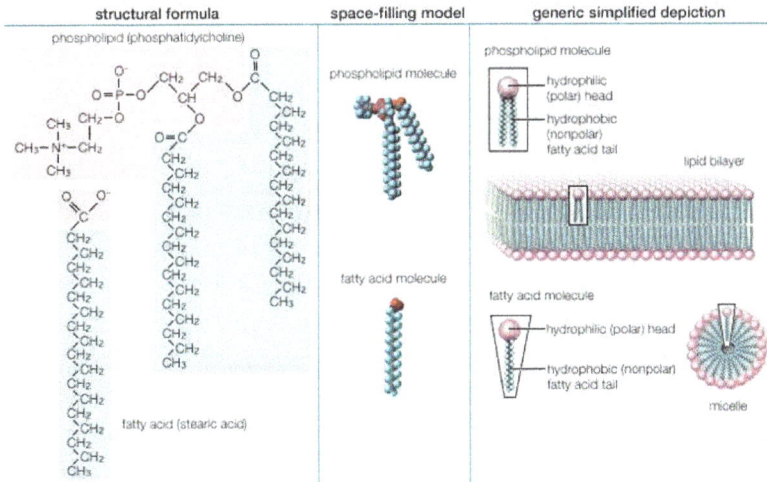

Structure and properties of two representative lipids both stearic acid (a fatty acid) and phosphatidylcholine (a phospholipid) are composed of chemical groups that form polar "heads" and nonpolar "tails." The polar heads are hydrophilic, or soluble in water, whereas the nonpolar tails are hydrophobic, or insoluble in water. Lipid molecules of this composition spontaneously form aggregate structures such as micelles and lipid bilayers, with their hydrophilic ends oriented toward the watery medium and their hydrophobic ends shielded from the water.

As soon as the biological catalysts became compartmentalized into small individual units, or cells, the units would have begun to compete with one another for the same resources. The active competition that ensued must have greatly accelerated evolutionary change, serving as a powerful force for the development of more efficient cells. In this way, cells eventually arose that contained new catalysts, enabling them to use simpler, more abundant precursor molecules for their growth. Because these cells were no longer dependent on preformed ingredients for their survival, they were able to spread far beyond the limited environments where the first primitive cells arose.

It is often assumed that the first cells appeared only after the development of a primitive form of protein synthesis. However, it is by no means certain that cells cannot exist without proteins, and it has been suggested that the first cells contained only RNA catalysts. In either case, protein molecules, with their chemically varied side chains, are more powerful catalysts than RNA molecules; therefore, as time passed, cells arose in which RNA served primarily as genetic material, being directly replicated in each generation and inherited by all progeny cells in order to specify proteins.

As cells became more complex, a need would have arisen for a stabler form of genetic information storage than that provided by RNA. DNA, related to RNA yet chemically stabler, probably appeared rather late in the evolutionary history of cells. Over a period of time, the genetic information in RNA sequences was transferred to DNA sequences, and the ability of RNA molecules to replicate directly was lost. It was only at this point that the central process of biology the synthesis, one after the other, of DNA, RNA, and protein appeared.

The Development of Metabolism

The first cells presumably resembled prokaryotic cells in lacking nuclei and functional internal compartments, or organelles. These early cells were also anaerobic (not requiring oxygen), deriving their energy from the fermentation of organic molecules that had previously accumulated on the Earth over long periods of time. Eventually, more sophisticated cells evolved that could carry out primitive forms of photosynthesis, in which light energy was harnessed by membrane-bound proteins to form organic molecules with energy-rich chemical bonds. A major turning point in the evolution of life was the development of photosynthesizing prokaryotes requiring only water as an electron donor and capable of producing molecular oxygen. The descendants of these prokaryotes, the blue-green algae (cyanobacteria), still exist as viable life-forms. Their ancestors prospered to such an extent that the atmosphere became rich in the oxygen they produced. The free availability of this oxygen in turn enabled other prokaryotes to evolve aerobic forms of metabolism that were much more efficient in the use of organic molecules as a source of food.

The switch to predominantly aerobic metabolism is thought to have occurred in bacteria approximately 2 billion years ago; about 1.5 billion years after the first cells had formed. Aerobic eukaryotic cells (cells containing nuclei and all the other organelles) probably appeared about 1.5 billion years ago, their lineage having branched off much earlier from that of the prokaryotes. Eukaryotic cells almost certainly became aerobic by engulfing aerobic prokaryotes, with which they lived in a symbiotic relationship. The mitochondria found in both animals and plants are the descendants of such prokaryotes. Later, in branches of the eukaryotic lineage leading to plants and algae, blue - green algae like organism was engulfed to perform photosynthesis. It is likely that over a long period of time these organisms became the chloroplasts.

The eukaryotic cell thus apparently arose as an amalgam of different cells, in the process becoming an efficient aerobic cell whose plasma membrane was freed from energy metabolism—one of the major functions of the cell membrane of prokaryotes. The eukaryotic cell membrane was therefore able to become specialized for cell-to-cell communication and cell signalling. It may be partly for this reason that eukaryotic cells were eventually more successful at forming complex multicellular organisms than their simpler prokaryotic relatives.

Evolutionary Physiology

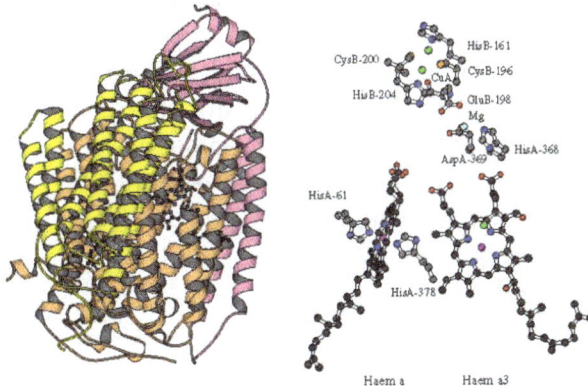

Evolutionary physiology represents an explicit fusion of two complementary approaches: evolution and physiology. Stimulated by four major intellectual and methodological developments (explicit consideration of diverse evolutionary mechanisms, phylogenetic approaches, incorporation of the perspectives and tools of evolutionary genetics and selection studies, and generalization of molecular techniques to exotic organisms), this field achieved prominence during the past decade. It addresses three major questions regarding physiological evolution:

(a) What are the historical, ecological, and phylogenetic patterns of physiological evolution?

(b) How important are and were each of the known evolutionary processes (natural selection, sexual selection, drift, constraint, genetic coupling/hitchhiking, and others) in engendering or limiting physiological evolution?

(c) How does the genotype, phenotype, physiological performance, and fitness interact in influencing one another's future values?

To answer these questions, evolutionary physiology examines extant and historical variation and diversity, standing genetic and phenotypic variability in populations, and past and on-going natural selection in the wild. Also, it manipulates genotypes, phenotypes, and environments of evolving populations in the laboratory and field. Thus, evolutionary physiology represents the infusion of paradigms, techniques, and approaches of evolutionary biology, genetics, and systematics into physiology.

Methods of Evolutionary Physiology

To understand what evolutionary physiology is, it is necessary to discuss the approaches which are required to study evolution of functions.

The study of function under extreme environmental conditions or in the presence of unusual factors in the milieu exterieur can reveal functional reserves and the range of Evolutionary plasticity. One example of the successful use of this approach is space physiology. Finally, substantial information on evolutionary physiology may be obtained in pharmacological and toxicological studies, because different sensitivity to the effects of toxic substances in animals of different classes can be found also in early stages of postnatal ontogenesis, particularly in birds and mammals. This makes it possible to analyze such phenomena as resistance, functional states of cell metabolic systems and their plasticity. All the above approaches of evolutionary physiology necessarily involve physiological, biochemical, biophysical, molecular biology, and morphological methods, as well as the methods of mathematical modelling and genetics.

The Evolution of Function

Evolutionary physiology differs from comparative and onto genetic physiology and other sciences modern physiology attributes increasing importance to investigations of membrane and molecular phenomena. While this provides for progress in contemporary physiology, there is a strong need to ascertain the place and importance of these molecular mechanisms in the activity of the organism as a whole. The concept of "function" is considerably broader than the activity of a whole organ, even when this organ is the main source of the function. For the example, salt and water balance provides convincing demonstration that the maintenance of a number of constants in the fluids of the milieu interior of Claude Bernard (e.g., constant volume, osmolality, ion concentrations) involves a number of organs and systems in vertebrates. In mammals, the kidney often serves as the main effectory organ of water and electrolyte homeostasis.

Evolution of Renal Function

The comparison of anatomical, functional, and biochemical specificity of the kidney in representatives of all the classes of vertebrates using the methods of evolutionary physiology makes it possible to formulate some regularities of the evolution of renal function.

The Increase of the Organ's Multifunctionality

The view that the kidney is a purely excretory organ by no means reflects the variety of homeostatic functions it fulfils. The kidney of Myxines is incapable of excreting hypotonic urine, whereas in lamprey this process is maintained quite efficiently. The osmoregulation function of the kidney is gradually established in the postnatal ontogenesis of mammals and it is disturbed as the result of a number of pathological processes. Data on the effect of chloride and sodium transport blockers (furosemide and amiloride) on the kidney in representatives of all classes of vertebrates, beginning with lamprey and extending to rat and human ontogenesis, illustrate the similarity of the membrane processes that underlie the mechanism of urine dilution. The kidneys of cyclostomata and fish do not respond to ADH by an increase in water reabsorption and thereby do not possess effective mechanisms of water economy. This function primarily originates in anuran amphibians; the kidney of homoiothermic vertebrates is the first to acquire the capacity to maintain water salt balance in conditions of extreme water deficit, due to its development of a capacity for osmotic concentration of urine. The mammalian kidney takes part in the regulation of blood volume, blood osmolality, ion concentration, pH, excretion of the end products of metabolism; the kidney produces physiologically active substances regulating arterial pressure, blood coagulation and modulating hormone effects, the active forms of vitamin D, the catabolism of peptide hormones takes place, and a number of other functions are executed. Multifunctionality of the kidney increases.

The study of renal function in postnatal mammalian ontogenesis demonstrates that the intensification of renal function reflects progressive renal evolution and is not a mere consequence of the acquisition of homoiothermic (as compared with poikilothermic vertebrates). At the initial stages of postnatal development, glomerular filtration rate increases markedly and the intensity of transport processes in tubules rises. Kidney of marine teleost have to re- move a large number of bivalent ions from blood, and reptiles and birds a large amount of uric acid. However, due to the low level of glomerular filtration rate, conditioned by comparatively low minute heart volumes in cold- blooded vertebrates the amount of the substances entering the kidney with the ultra-filtrate is lower than the amount to be excreted. This requirement makes the occurrence of an additional blood supply to the kidney necessary. In the evolution of certain classes of vertebrates another additional system of renal blood supply evolved. In the majority of vertebrates, with the exception of cyclostomates, freshwater fish and mammals, there exists a Reno portal system through which the venous blood from the back part of the body flows towards the kidneys. This sharply increases the potential capabilities of the kidney to excrete a number of substances by secretion into the tubule lumen. Hence, an increase of the glomerular filtration rate and thereby a rise in the entry of various substances in the nephron lumen and a reabsorption of a part of them provides for the excretion of necessary substances.

Substitution of an Organ or its Function

Adaptation to new environmental factors and to new conditions of existence may lead to a change in the function of an organ and the fulfilment of quite different activity by it. Data on water–salt balance give evidence that the transformation of the function of an organ or the replacement of one organ by another in maintaining a new function might be determined not only by the potential capacities of the initial structure, but also by the level of development of the organism's other functional systems. Vertebrate adaptation to sea water, except in mammals, is associated with the participation of a number of organs in hypo osmotic regulation. To achieve osmoregulation, some animals drink sea water and produce fresh water by sodium and chloride ions secreting through the gill (in teleost), rectal glands (in elasmobranch), or salt glands (in reptiles and birds). The kidneys of these animals cannot yet maintain a sufficiently high level of urine osmotic concentration and water economy; these animals do not have the counter-current multiplying system of the renal medulla which develops only in birds and in mammals. Thus, in adapting to a hyper- osmotic environment (the ocean), the functional substitution in osmoregulation of the kidney by gill cells or by salt glands takes place. Only in mammals is this function fulfilled by the kidney alone.

Regression of Function

The migration of teleost from fresh water to the sea led to the alteration of the morpho functional organization of their kidneys. In marine fish, the kidneys do not in fact participate in osmoregulation, but usually excrete urine almost isosmotic with blood plasma; the excretion of slightly hypotonic urine has been reported in only a few species. The function of the kidneys in these fish consists primarily of the excretion of surplus bivalent ions (magnesium ions, sulphates and some others) that enter the organism in large quantities when it swallows sea water. The essential change in the kidney structure of these fish, illustrated in many marine teleost species, is the reduction of the glomerular apparatus; in several species, the kidneys completely lack glomeruli, and a number of species of these fish have also lost the distal tubules. This regression of renal function, manifested by its inability to excrete diluted urine, favoured the adaptation to the sea water. It led to a transformation of the kidney that enhanced its potential for excreting bivalent ions and decreasing the loss of water with urine. The basis of such a transformation could be the utilization of initial forms of renal cell activity: The secretion of Mg and Ca ions is probably ensured by ion-exchange processes of the type Na/Mg or Na/Ca.

Irreversibility of the Regressive Evolution of Function

The permanently increasing differentiation that results from the adaptation to a narrow ecological niche entails a decrease in plasticity. The acquisition of progressive characters may be accompanied by the loss of other properties that cannot then reappear in the same form. In palaeontology, Dollo expressed the idea about the irreversibility of the evolution. The degradation of structures in certain groups of animals is irreversible; although evolution will continue by changing its direction and following another possibly progressive path, the new pathway will be altogether different. The close connection between structure and function expresses itself when, in animals possessing still higher specialization with respect to particular conditions of existence, increasing adaptation to an environment may be accompanied by the loss of these or other functions during the adaptation of many generations. For example, wild rodents and the white rat excrete the surplus water equally effectively. The necessity of the water economy under water deficit conditions in the desert has resulted in the adaptive evolution of the big gerbil as com- pared to the water vole living near the fresh water. The kidneys of these both rodent species not only acquired a number of features favouring adaptation to their habitats but also lost certain functions. In the kidney of the water vole, the renal papilla was reduced, nephrons with a long Henle loop disappeared, and the ion content of the renal medulla decreased. The kidneys of the water vole lost the capacity to produce urine of high osmolality. Meanwhile, the big gerbil, which is an inhabitant of deserts, can manage without drinking water at all. This is ensured by the exceptional development of long thin Henle loops and of changes in the renal inner medulla which makes the kidney capable of excreting urine of osmolality that is more than 15 times higher than that of blood. Even with these changes of the osmoregulating function of the kidney its other functions, such as volume and ion regulation, remained equally efficient.

Evolution of Functional Units

The morpho functional unit of the vertebrate kidney is the nephron. Considering the specifics of nephron structure, the presence or the absence of glomeruli and certain parts of tubules, and the changes of their functional properties in representatives of different classes of vertebrates, as well as in ontogenesis and in pathology, we may suggest the following principles of the evolution of their functions.

Increase of Heterogeneity

This occurs in the form of appearance of several nephron populations in the kidney The heterogeneity of nephrons was observed in the kidneys of representatives of every class of vertebrates. The functional importance of this heterogeneity is clearly revealed in birds and mammals, where various nephron populations play different roles in the osmotic concentration of urine. Secretion of vasopressin is necessary for the increase of heterogeneity of nephrons in postnatal ontogenesis. In adult rats with hereditary diabetes insipidus (Brattleboro rats) in which the secretion of vasopressin is absent, superficial nephrons prevail over juxtamedullar ones, contrary to the pattern in normal rats. Injections of vasopressin to Brattleboro rats beginning at 2 weeks of age restore the normal nephron heterogeneity in their kidneys. These results indicate that genotypic and phenotypic factors play a role in the emergence of the heterogeneity of nephron populations. Osmotic concentration in mammalian kidneys with a long Henle loop is conditioned by the accumulation of sodium ions and urea: in the renal medulla in birds and in the outer medulla of mammalian kidney, this process is maintained mainly by sodium chloride. Structural nephron heterogeneity is closely connected to functional and biochemical distinctions.

Increase of Differentiation

The glomeruli and the proximal and distal segments of the nephron are found in the kidneys of all classes of vertebrates, from cyclostomates (lamprey) to mammals except the cases described above the increase of the differentiation of distal segments of nephron into an increased number of parts in the formation of the nephron loop is clearly revealed in homoiotherms. The differentiation occurs in the nephrons of the elasmo branch and is probably necessary for reabsorption of urea The differentiation involves the various nephron elements: cells, intercellular contacts, and the connective tissue surrounding them. In a number of cases, as, for example, in some marine teleosts, the number of parts in nephrons is reduced, and glomeruli or distal tubules may be absent. The tendency toward increasing differentiation of nephrons in the phylogenesis of vertebrates and in ontogenesis of mammals, and the dedifferentiation of nephrons in pathology, support the suggestion that the progressing differentiation is one of the peculiarities of the evolution of functional units. This process involves the alteration of their structural, functional, biochemical, and biophysical characteristics.

Increase of Reabsorption and Secretion

One specific direction of renal evolution, which increases in the reabsorption of ultra-filtrate, is reflected in the activity of nephrons. In the proximal convoluted tubule of mammals, the rate of reabsorption of fluid, calculated per unit of tubule length, is higher than in poikilothermal vertebrates; this rate also is increased in mammalian ontogenesis. In contrast, the opposite occurs in pathology (e.g., with chronic renal failure). The said above concerns the increase of reabsorption of the main bulk of the filtered substances. Meanwhile, when it is necessary to remove certain substances.

The Formation of Morph Functional Complexes

The activity of nephrons is closely related to the vascular system of the kidney. The interstitium plays an important role in a number of renal functions. One sign of the progres sive evolution of the kidney is the formation of the system for the osmotic concentration of urine. This function is fulfilled by the complex of the Henle loop with straight vessels. Together with the highly differentiated interstitium, they form the medulla of the kidney: a counter- current multiplication system. An integrated concept of the evolution of functional units in the kidney is created only when the interconnections of blood and lymph vessels, nervous elements, and connective tissues in the development of nephrons is understood.

Increase of the Capacity to Regulate Functional Activity

The increase in the differentiation of nephrons, their close connection with the vascular system, and the higher degree of regulation of functions at each level of the nephron under the influence of physiologically active substances create better possibilities for the adaptation of renal function to the requirements of the organism. In the nephron, disturbances in the system of reabsorption of various ions in the proximal segment may be corrected in the distant parts of the nephron. This pattern is most obviously revealed in mammals, where the decrease of proximal reabsorption of Na and Cl ions may be accompanied by their strengthened compensatory increased reabsorption in the ascending limb of the Henle loop. It is quite possible that the increase of glomerular filtration rate and proximal reabsorption is an important evolutionary acquisition of homoiothermal animals. In combination with the formation the thick ascending limb of the Henle loop, this forms a powerful system of reabsorption of a number of ions. This system adapted to a wide variability in amounts of reabsorbed substances which can compensate the alteration of their transport in the proximal parts of the nephron.

Evolution of Functions of Specialized Cells

In embryogenesis, the nephron cells of vertebrates is developed from different initial groups of cells; in cyclostomates fishes, and amphibians, the definitive kidney by its origin is related to mesonephros whereas in reptiles, birds and mammals to metanephros. The nephron cell like any other of the organism's cells, must possess a complete set of molecular systems that maintain the transport of various organic and inorganic substances necessary for its life activity. Comparison of the data on physiological, biochemical, and ultrastructural peculiarities of cells from different parts of a nephron will permit the evaluation of the ways and means by which the initial cell forms are transformed. Analyses of how nephron cells at different sites become specialized and comparisons of corresponding parts of the nephron in vertebrate animals at different levels of individual development and from different systematic groups provide much information that will help us to judge about the regularities of the evolution of specialized cells.

Formation of Asymmetrical Cell

In the evolution of nephron cells, as well as that of the cells of any other osmoregulatory and excretory organ, an important role is played by specializations of its apical and basal membranes, which determine the formation of the asymmetrical cell and thereby allow directed transport. It is important for the reabsorption and secretion of substances that ion channels and pumps are located on the opposite plasma membranes of the cell. This provides for transport of sub stances from the tubule lumen into the blood. As to for secretion, it requires another, opposite organization of ion channels, pumps, and transporters. In evolution, the specialization of cells provides such redistributions of the initial molecular transport mechanisms so that they may actually disappear from one plasma membrane and persist in an- other. The specialization of transport systems can maintain secretion from the blood into the tubule lumen not only of endogenic substances but also of xenobiotic, organic acids, and bases.

Formation of Specialized Ultrastructure's

The different properties of the apical and basolateral nephron cell membranes are the basis for the capacity of these cells for directed transport of substances in the course of their reabsorption and secretion. The properties of the apical cell membrane are not the same in different parts of the nephron. In the proximal segment of the nephron the cells possess numerous microvilli forming the brush border; in the later parts of the nephron only separate microvilli are seen. Functionally, the apical membrane of the cells of a proximal tubule is capable not only of reabsorption of ions from the tubular fluid, but also of transport of glucose, amino acids, and proteins into the cell for their subsequent reabsorption into the blood. In other nephron parts (with the exception of collecting ducts) in apical plasma membranes of the cells, these molecular devices are practically absent, at these sites the reabsorption of organic substance does not take place in substantial quantities; instead, ions alone are reabsorbed.

Glucose and amino acids can enter the cytoplasm of the cells from the extracellular fluid through the basolateral membranes. In the cells of the proximal tubule there are numerous pinocytosis vesicles, providing the capture and subsequent hydrolysis of protein for its reabsorption. The fundamental ultrastructural distinctions among different nephron parts, together with the different levels of their enzyme activity and differences in the systems of transport of various substances, indicate the role of the differentiation of plasma membranes as well as a number and location of organelles in the formation of specialized cells in different parts of nephrons.

Increase of Intensity of Reabsorption and Secretion

Both the mass of the kidney relative to the size of an animal, and the relative length of proximal tubule of vertebrates remain roughly constant, so the higher efficiency of the tubules work is accomplished by the intensification of cell activity. This expresses itself in a larger number of mitochondria, an increase of the number of cristae they contain, an increase in oxygen consumption, and infoldings of the basal plasma membrane and hence a larger surface area where Na, K-ATPase is localized, contributes to an increase of active ion transport.

Changes in a Number of Organelles

Among the factors that maintain the intensive work of the cell, an important part is played by the increase in the number of organ elles: the "polymerization" or organelles, corresponding to one of the principles of protozoan cell evolution put forward by G.I. Polyansky. Attention must be paid to the fact that in the course of its specialization within an organ of a metazoan organism cell evolution does not necessarily follow the route of intensification of the metabolism. The differentiation of the mammalian nephron cell in the formation of the thin limb of Henle loop provides an example in which the formation of flattened epithelial cells poor in mitochondria and lacking the brush border, with scarce folds on the basolateral membrane, can be observed. The level of glycolysis in these cells is high because they function in the inner medulla of the kidney where in some species the osmotic pressure under the conditions of osmotic concentration may exceed 5000 mOsm/kg H_2O. The progressive evolution of an organ may be accompanied by the changes in certain types of its cells, where the fulfilment of new functions requires the simplification of a number of aspects of cell organization, the decrease of the number ("oligomerization") of their organelles, and decreases in intensity of metabolism. The clearly expressed differentiation of the nephron cell, changes in its function and in the structure of cells are manifested by the increase of individual forms of the initial activity of the cell. Thus, any cell, in order to fulfil its functions, to construct its own components and to cover its own energetic expenditures requires input from the extracellular fluid of amino acids, glucose, free fatty acids, and must contain Na-dependent carriers for the transport of these substances. In most of the cells of renal tubules this process occurs just as in other cells of the body, but it is only in one segment of the nephron that this form of cellular activity is transformed into a specialized transport function. The increase of this initial cell activity reaches such a high intensity that in human kidney the reabsorption of glucose alone amounts to 990 mmole per day (for comparison it is enough to say that 1 L of blood plasma contains only 5.5 mmole of glucose). In these cells directed transport from the tubule lumen into the blood provides for the reabsorption of large quantities of amino acids, vitamins, and in fact, all other vitally necessary substances.

The study of the morphological and the functional evolution of a specialized cell in a multicellular organism, unlike that of protozoans, must take into account the functional organization of the cell contacts. Many investigations emphasize a high degree of specialization of cell contacts (the tight junction, desmosomes, and gap junctions). In the proximal tubule of a mammalian kidney the cell junctions are highly permeable for certain ions, but does not let many non-electrolytes pass through, its electrical resistance being some 5 Ohm cm². In the terminal parts of the distal nephrons of the same animals where the electrical resistance amounts to hundreds and sometimes even thousands Ohm cm² the tubule wall lets neither ions nor electrolytes pass; the ultrastructure of tight junctions becomes quite different also. In amphibians, where the proximal tubule wall is less permeable for the ions and the level of reabsorption is lower than in mammals, the electrical resistance reaches 50 Ohm cm². Consequently, a significant component of evolution at the cellular level of the organization of multicellular systems is in the specialization of cell junctions.

Not only the specific forms of its cellular activity that are important to an organism, but also its capacities for regulation depending on its state and the constancy of the milieu interieur that is ensured by these capacities. Efferent nerves participate in regulation of the nephron cells' function, and various hormones and autacoids of higher animals have been found in animals of even of the most primitive organization; judging by the available data, in evolution of the kidney, the number of regulatory factors increases and their effects change. This concerns the action of arginin vasotocine on the permeability of tubules to water which appeared only from the amphibia and the appearance of the response of Care absorbing systems to hormones. The development of specific responses to these hormones is determined by the formation of several mechanisms, but it depends primarily on presence of specific receptors in cells of the renal tubules; fine regulation is due to the formation of autacoids and their impact on cells.

One important parameter of the functional state of a cell is its response to physiologically active substances. Consider, for example, the effect on the kidney of cisplatin, an anti-tumour drug with a nephrotoxic effect. After administration of equal doses, there is more damage to the cells of proximal tubules in rats and pigeons, than in amphibia and teleost. Thus, one must conclude that the evolution of nephron cells was accompanied by the change of sensibility to physiologically active substances. Thereby knowledge of the ways in which the functions of organs were changed in evolution must inevitably be based on clarifying the regularities of the evolution of the functions of specialized cells.

Functional Evolution

The term "functional evolution" is often regarded as the antithesis of morphological evolution. In reality, the scientific implications of this term are far deeper.

The problems of functional evolution remained insufficiently elaborated. Functional evolution is one of the most complicated theoretical parts of evolutionary physiology as its goal is to give a final account of a number of the most important problems of this science. From our viewpoint, the problem of functional evolution without doubt is essential for the physiological approach towards investigating the evolutionary process. Among functional evolution problems are: the problem of the physico-chemical bases that determine the development of functions, the problem of the development of the integrity of the organism, the question of the origin and mechanisms of an organism's adaptation to the environment, as well as some other problems that are undoubtedly important for the evolutionary theory but cannot be solved through analyzing the evolution of separate functions.

Physico-chemical Factors in the Evolution of Function

The study of a broad spectrum of species has shown that, irrespective of the diversity of the forms of functional organization of the organs investigated (e.g., respiratory, digestive, excretory and sensory organs), the plan of their structure and activity is uniform across animals of different lines. The application of physiological methods allows us to analyze why similar organs emerged independently in evolution but had a similar functional organization. On the basis of the concepts in general physiology and physico-chemical biology, we can attempt to answer questions about the "permissible" and "forbidden" ways in which a given function can develop. In the middle of the 1960s, we suggested that one approach to solve the problem of functional evolution is to analyze the molecular and membrane mechanisms of the function under study, because these mechanisms predetermine the particular (such rather than another) way of the organization of each physiological system.

It is convenient to illustrate this statement using several examples. The functional organization of organs that participate in water–salt homeostasis is similar to that of glands of exocrine secretion. In evolution, the combination of ultrafiltration or its equivalent with the sub sequent reabsorption and secretion of substances in tubules is consistently observed. It occurs in the protonephridia of platyhelmintes, rotatoria, nemertini, and priapulidae, in the metanephridia of annelids, in coelomoducts, for example, the Bojanus organ of molluscs, the antennal glands of crustaceans, the kidney of vertebrates, and in the salivary glands of mammals. Since secretions and excretions are based on the work of these organs, it might be thought that the two stage process of urine formation and secretion in these organs predetermines the membrane and molecular mechanisms of water transport (which is a passive transport in all known cases of osmoregulating reactions) and sodium transport (which, contrary to water, may also be an active transport by means of a sodium pump). As the result of reabsorption through the duct or tubule wall, which is impermeable to water, the formation of osmotically free water takes place. However, some glands (e.g., salt glands) work on the basis of fluid secretion without the primary process of ultrafiltration; in these cases, hypertonic fluid is formed.

The molecular mechanisms of transport of water and sodium probably predetermine not only the functional organization of the kidney but also that of the whole system of water-salt balance. In homoiosmotic animals adapted to sea water, several organs can be involved in this process. To freshen the water, marine teleost, reptiles, and birds drink sea water and obtain osmotically free water through the active excretion of sodium salts by gill cells and salt glands. Another type of adaptation, in marine elasmobranchia and amphibians, is through sharp increases in the concentration of urea and trimethylamine oxide in body fluids, which produce a passive inflow of water into the body through the integuments along the osmotic gradient. The direct secretion of water itself might be an alternative, but this is so dissipative of energy that living beings have evolved instead to use the two mechanisms described above. With respect to the functional evolution of water salt balance, therefore, we have apparently succeeded in explaining the given pathways of development: based on by the membrane peculiarities of water and sodium transport.

This is only one aspect of the problems pertinent to functional evolution. Biological phenomena, especially those related to the evolution of the primary vital processes, may not be wholly reduced to simplified physicochemical concepts. In this connection, Einstein's words of 1953 come to mind; he said that physics first affects the development of medicine by making people believe in scientific method, however, physics tempted biologists to interpret life pro- cesses too primitively. We considered above the importance of molecular and membrane mechanisms of water and sodium ion transport in determining the anatomical and functional organization of the kidney and glands of external secretion.

However, another evolutionary variant is possible when the anatomical specificity of an organ serves as prerequisites for the development of a novel molecular composition of the milieu interieur. In the larvae of higher insects with potassium hemolymph, in which the concentration of sodium is very low, the Na/K ratio in cells appears the same as in animals of other systematic groups. In order to provide for the cell volume regulation, they accumulate a high concentration of amino acids in their blood and they reduce the hemolymph concentration of divalent cations by binding them not only with hemolymph protein but also with organic anions. The evolutionary emergence of a milieu interieur so unusual in its composition was determined, in particular, by the anatomical specificity of the excretory organs of insects. Malpighian vessels lack the structures for the ultrafiltration of fluids that are present in the majority of cases in coelomoducts and protonephridia and metanephridia. As a result of this lack, amino acids and organic acids, which in other animals are permanently filtered and reabsorbed, are retained in the hemo lymph. Malpighian vessels are also present in the lower in sects, which show the same chemical composition of hemolymph as most other animals, and later evidently were utilized as a "preadaptation" in the evolutionary development of higher insects.

In vertebrates and in the majority of invertebrates in whose blood sodium is dominant, sodium takes part in the transmembraneous transfer of many ions and organic sub stances through the mechanism of sodium-dependent secondary active transport. An example of another variant of the molecular mechanism of ion-dependent transport may be the hemolymph of the higher insects in which potassium and not sodium usually appears to be the dominant cation. A potassium-dependent rather than a sodium dependent transport of organic and inorganic substances has been revealed in the plasma membranes in higher insects. This aspect of the study of functional evolution deals with the essence of physiological phenomena that occur in living beings of different levels of organization. In any given case, we actually consider both transient and intransient phenomena. The mechanism of a physiological phenomenon is carried in the transient generations of living beings, which supersede each other while revealing newly appearing organs and processes in different phyletic lines; the complication in understanding this aspect of functional evolution is that, as soon as the mechanism of a given sys- tem becomes uniform, so that there are no differences with regard to its principles in lower and higher forms, there seems to be no development. But the differences consist not in the principles, so far as they have been discovered, but in the form of their realization, i.e., just in what is already relevant to the problem of the evolution of function. This idea may be expressed in other words. The essence of things cannot evolve, but only the extent of our penetration into the depths of phenomena, which depends on our ability to capture the essence of the physiological processes that can undergo change. The study of functional evolution is com mitted to understanding how the nature of phenomena and physiological phenomena in particular, determines the ways in which biological functions have developed and continue to develop.

Functional Specialization of Cells

The problem of the origin of functions, the mechanisms by which cells become specialized, constitutes one of the important aspects of functional evolution. Chemical evolution led to the emergence of the primary forms of life, and an enormous period of time (more than 2 billion years of the 4.6 billion years of the existence of the Earth, according to modern estimates), was required to build the eukaryote cell.

References

- Evolutionary-biology: biology-online.org, Retrieved 27 April 2018

- Biological-evolution, biology-and-genetics/biology-general, science-and-technology: encyclopedia.com, Retrieved 18 July 2018

- Evolutionary-dynamics: bsse.ethz.ch, Retrieved 31 March 2018

- Unifying: ucl.ac.uk, Retrieved 11 March 2018

- The-process-of-differentiation-37470: britannica.com, Retrieved 11 May 2018

Chapter 3

Natural Selection

Natural selection is the preferential survival and reproduction of specific individuals based on differences in their heritable traits. The aim of this chapter is to explore the fundamentals of natural selection, such as heritable traits, sexual selection, stabilizing selection, directional selection, negative selection, gene selection, microevolution and macroevolution, among others.

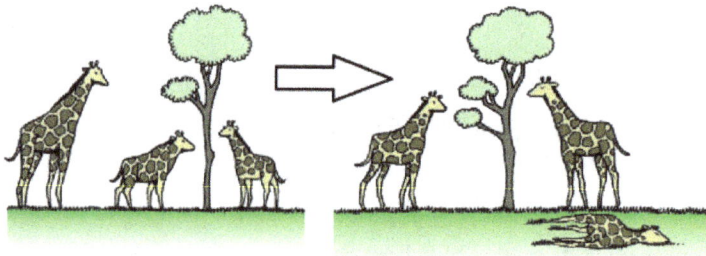

Natural Selection in action

Natural selection is the engine that drives evolution. The organisms best suited to survive in their particular circumstances have a greater chance of passing their traits on to the next generation. But plants and animals interact in very complex ways with other organisms and their environment.

Natural Selection Examples

Here are examples of natural selection:

- In a habitat there are red bugs and green bugs. The birds prefer the taste of the red bugs, so soon there are many green bugs and few red bugs. The green bugs reproduce and make more green bugs and eventually there are no more red bugs.

- In an ecosystem, some giraffes have long necks and others have short ones. If something caused low-lying shrubs to die out, the giraffes with short necks would not get enough food. After a few generations, all the giraffes would have long necks.

- A species of rats live in a certain type of tree with the branches evenly spaced. Smaller rats could not reach from branch to branch and larger rats would break the branches and fall. Soon, all rats were just the right size for the tree branches.

- Deer mice that migrated to the sand hills of Nebraska changed from dark brown to light brown to better hide from predators in the sand.

- Galapagos finches all have different types of beaks. During drought, the finches with the larger beaks survived better than those with smaller beaks. During rainy times, more small seeds were produced and the finches with smaller beaks fared better.

- In one ecosystem, lizards that had long legs could climb better to avoid floods and reach food.

- Insects become resistant to pesticides very quickly, sometime in one generation. If an insect is resistant to the chemical, most of the offspring will also be resistant. Considering that insect generations can be a matter of weeks, insects in an area can become immune to a chemical within months.

- The bacterium Pseudomonas metabolizes nylon; but, when a certain type of this bacterium that did not eat nylon was placed in an environment where nylon was the only food, the bacterium evolved until it ate the nylon.

- Peacock females pick their mate according to the male's tail. The ones with the largest and brightest tails mate more often. Today, males that do not have bright feathers are very rare.

- The peppered moth used to be a light color with black spots. When the atmosphere in London became filled with soot because of the Industrial Revolution, the white trees became darker and light colored moths were eaten by birds more readily. Within months, moths became darker and lighter moths were rare. After the Industrial Revolution, light colored moths were found in greater numbers.

- Rat snakes are all very similar except in coloring. Some are striped and some are green, black, and orange. This is due to living in many different types of terrain and adapting to the environment.

- Warrior ants have a chemical signal that tells other ants in the family not to attack. Some have adapted and learned to imitate the chemical signal from other colonies so they can invade and take over another colony and the workers will never know.

- Sharks are colored white on the underside and blue or gray on the top. This is their camouflage as the top blends with the water color to someone looking down into the water and the bottom blends with the light coming through the water from above.

- Bacteria have become resistant to antibiotics and this can happen very quickly, since bacteria can produce several generations within one day. The strongest bacteria are the last to die and the antibiotics sometimes do not kill all of them.

- The field mustard plant survived a drought in southern California because of genetic changes that made it produces flowers earlier. Plants that survive a drought have to have a short growing season and the field mustard evolved quickly.

- Because of its long body, the moray eel's mouth did not produce enough suction to catch prey; so, it adapted and grew a second set of jaws and teeth.

- Humans have evolved and are still evolving. There are people more resistant to malaria that live in Africa.

- Humans don't become lactose intolerant as some species do. This is thought to be because of the domestication of cattle.

- Early humans survived because of a certain hand shape that could either toss a spear or throw a rock the best.

- In the 18th century, researchers found in an island community where there was a genetic disposition to mate earlier and have larger families. Over 140 years, the age at first reproduction dropped from 26 to 22 and this was due to natural selection.

Charles Darwin coined the term "natural selection. Darwin's concept of natural selection was based on several key observations:

- Traits are often heritable. In living organisms, many characteristics are inherited, or passed from parent to offspring. (Darwin knew this was the case, even though he did not know that traits were inherited via genes).

- More offspring are produced than can survive. Organisms are capable of producing more offspring than their environments can support. Thus, there is competition for limited resources in each generation.

- Offspring vary in their heritable traits. The offspring in any generation will be slightly different from one another in their traits (color, size, shape, etc.), and many of these features will be heritable.

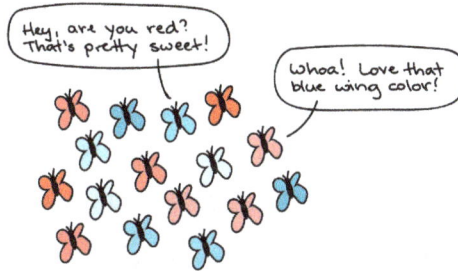

Based on these simple observations, Darwin concluded the following:

- In a population, some individuals will have inherited traits that help them survive and reproduce (given the conditions of the environment, such as the predators and food sources present). The individuals with the helpful traits will leave more offspring in the next generation than their peers, since the traits make them more effective at surviving and reproducing.

- Because the helpful traits are heritable, and because organisms with these traits leave more offspring, the traits will tend to become more common (present in a larger fraction of the population) in the next generation.

- Over generations, the population will become adapted to its environment (as individuals with traits helpful in that environment have consistently greater reproductive success than their peers).

Darwin's model of evolution by natural selection allowed him to explain the patterns he had seen during his travels. For instance, if the Galápagos finch species shared a common ancestor, it made sense that they should broadly resemble one another (and mainland finches, who likely shared that common ancestor). If groups of finches had been isolated on separate islands for many generations, however, each group would have been exposed to a different environment in which different heritable traits might have been favored, such as different sizes and shapes of beaks for using different food sources. These factors could have led to the formation of distinct species on each island.

Mechanism

Heritable Variation, Differential Reproduction

Natural variation occurs among the individuals of any population of organisms. Some differences may improve an individual's chances of surviving and reproducing such that its lifetime reproductive rate is increased, which means that it leaves more offspring. If the traits that give these individuals a reproductive advantage are also heritable, that is, passed from parent to offspring, then there will be differential reproduction, that is, a slightly higher proportion of fast rabbits or efficient algae in the next generation. Even if the reproductive advantage is very slight, over many generations any advantageous heritable trait becomes dominant in the population. In this way the natural environment of an organism "selects for" traits that confer a reproductive advantage, causing evolutionary change, as Darwin described. This gives the appearance of purpose, but in natural selection there is no intentional choice. Artificial selection is purposive where natural selection is not, though biologists often use teleological language to describe it.

During the industrial revolution, pollution killed many lichens, leaving tree trunks dark. A dark (melanic) morph of the moth largely replaced the formerly usual light morph (both shown here). Since the moths are subject to predation by birds hunting by sight, the color change offers better camouflage against the changed background, suggesting natural selection at work.

The peppered moth exists in both light and dark colours in Great Britain, but during the industrial revolution, many of the trees on which the moths rested became blackened by soot, giving the dark-coloured moths an advantage in hiding from predators. This gave dark-coloured moths a better chance of surviving to produce dark-coloured offspring, and in just fifty years from the first dark moth being caught, nearly all of the moths in industrial Manchester were dark. The balance was reversed by the effect of the Clean Air Act 1956, and the dark moths became rare again, demonstrating the influence of natural selection on peppered moth evolution. A recent study, using image analysis and avian vision models, shows that pale individuals more closely match lichen backgrounds than dark morphs and for the first time quantifies the camouflage of moths to predation risk.

Fitness

The concept of fitness is central to natural selection. In broad terms, individuals that are more "fit" have better potential for survival, as in the well-known phrase "survival of the fittest", but the precise meaning of the term is much more subtle. Modern evolutionary theory defines fitness not by how long an organism lives, but by how successful it is at reproducing. If an organism lives half as long as others of its species, but has twice as many offspring surviving to adulthood, its genes become more common in the adult population of the next generation. Though natural selection acts on individuals, the effects of chance mean that fitness can only really be defined "on average" for the individuals within a population. The fitness of a particular genotype corresponds to the average effect on all individuals with that genotype.

Competition

In biology, competition is an interaction between organisms in which the fitness of one is lowered by the presence of another. This may be because both rely on a limited supply of a resource such as food, water, or territory. Competition may be within or between species, and may be direct or indirect. Species less suited to compete should in theory either adapt or die out, since competition plays a powerful role in natural selection, but according to the "room to roam" theory it may be less important than expansion among larger clades.

Competition is modeled by r/K selection theory, which is based on Robert MacArthur and E. O. Wilson's work on island biogeography. In this theory, selective pressures drive evolution in one of two stereotyped directions: r- or K-selection. These terms, r and K, can be illustrated in a logistic model of population dynamics:

$$\frac{dN}{dt} = rN\left(1 - \frac{N}{K}\right)$$

Where r is the growth rate of the population (N), and K is the carrying capacity of its local environmental setting. Typically, r-selected species exploit empty niches, and produce many offspring, each with a relatively low probability of surviving to adulthood. In contrast, K-selected species are strong competitors in crowded niches, and invest more heavily in much fewer offspring, each with a relatively high probability of surviving to adulthood.

Types of Selection

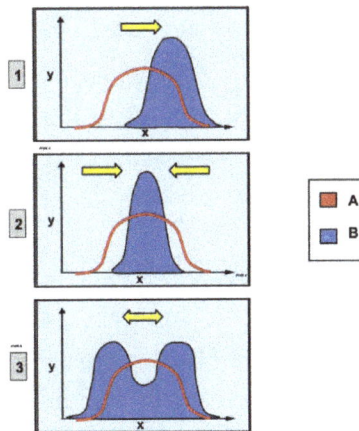

1: directional selection: a single extreme phenotype favoured. 2, stabilizing selection: intermediate favoured over extremes. 3: disruptive selection: extremes favoured over intermediate. X-axis: phenotypic trait, Y-axis: number of organisms, Group A: original population, Group B: after selection

Natural selection can act on any heritable phenotypic trait, and selective pressure can be produced by any aspect of the environment, including sexual selection and competition with members of the same or other species. However, this does not imply that natural selection is always directional and results in adaptive evolution; natural selection often results in the maintenance of the status quo by eliminating less fit variants.

Selection can be classified in several different ways, such as by its effect on a trait, on genetic diversity, by the life cycle stage where it acts, by the unit of selection, or by the resource being competed for.

Selection has different effects on traits. Stabilizing selection acts to hold a trait at a stable optimum, and in the simplest case all deviations from this optimum are selectively disadvantageous. Directional selection favours extreme values of a trait. The uncommon disruptive selection also acts during transition periods when the current mode is sub-optimal, but alters the trait in more than one direction. In particular, if the trait is quantitative and univariate then both higher and lower trait levels are favoured. Disruptive selection can be a precursor to speciation.

Alternatively, selection can be divided according to its effect on genetic diversity. Purifying or negative selection acts to remove genetic variation from the population (and is opposed by *de novo* mutation, which introduces new variation.In contrast, balancing selection acts to maintain genetic variation in a population, even in the absence of *de novo* mutation, by negative frequency-dependent selection. One mechanism for this is heterozygote advantage, where individuals with two different alleles have a selective advantage over individuals with just one allele. The polymorphism at the human ABO blood group locus has been explained in this way.

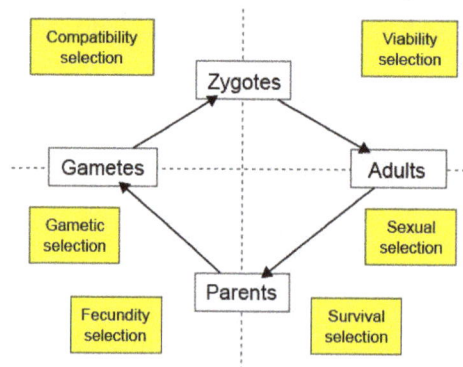

Different types of selection act at each life cycle stage of a sexually reproducing organism.

Another option is to classify selection by the life cycle stage at which it acts. Some biologists recognize just two types: viability (or survival) selection, which acts to increase an organism's probability of survival, and fecundity (or fertility or reproductive) selection, which acts to increase the rate of reproduction, given survival. Others split the life cycle into further components of selection. Thus viability and survival selection may be defined separately and respectively as acting to improve the probability of survival before and after reproductive age is reached, while fecundity selection may be split into additional sub-components including sexual selection, gametic selection, acting on gamete survival, and compatibility selection, acting on zygote formation.

Selection can also be classified by the level or unit of selection. Individual selection acts on the individual, in the sense that adaptations are "for" the benefit of the individual, and result from selection among individuals. Gene selection acts directly at the level of the gene. In kin selection and intragenomic conflict, gene-level selection provides a more apt explanation of the underlying process. Group selection, if it occurs, acts on groups of organisms, on the assumption that groups replicate and mutate in an analogous way to genes and individuals. There is an ongoing debate over the degree to which group selection occurs in nature.

Finally, selection can be classified according to the resource being competed for. Sexual selection results from competition for mates. Sexual selection typically proceeds via fecundity selection, sometimes at the expense of viability. Ecological selection is natural selection via any means other than sexual selection, such as kin selection, competition, and infanticide. Following Darwin, natural selection is sometimes defined as ecological selection, in which case sexual selection is considered a separate mechanism.

Sexual Selection

The peacock's elaborate plumage is mentioned by Darwin as an example of sexual selection, and is a classic example of Fisherian runaway, driven to its conspicuous size and coloration through mate choice by females over many generations.

Sexual selection as first articulated by Darwin (using the example of the peacock's tail) refers specifically to competition for mates, which can be *intrasexual*, between individuals of the same sex, that is male–male competition, or *intersexual*, where one gender chooses mates, most often with males displaying and females choosing. However, in some species, mate choice is primarily by males, as in some fishes of the family Syngnathidae.

Phenotypic traits can be displayed in one sex and desired in the other sex, causing a positive feedback loop called a Fisherian runaway, for example, the extravagant plumage of some male birds such as the peacock. An alternate theory proposed by the same Ronald Fisher in 1930 is the sexy son hypothesis, that mothers want promiscuous sons to give them large numbers of grandchildren and so choose promiscuous fathers for their children. Aggression between members of the same sex is sometimes associated with very distinctive features, such as the antlers of stags, which are used in combat with other stags. More generally, intrasexual selection is often associated with sexual dimorphism, including differences in body size between males and females of a species.

Natural Selection in Action

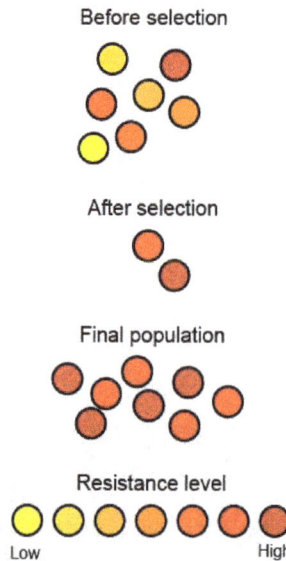

Before selection

After selection

Final population

Resistance level

Low   High

Selection in action: resistance to antibiotics grows though the survival of individuals less affected by the antibiotic. Their offspring inherit the resistance.

Natural selection is seen in action in the development of antibiotic resistance in microorganisms. Since the discovery of penicillinin 1928, antibiotics have been used to fight bacterial diseases. The widespread misuse of antibiotics has selected for microbial resistance to antibiotics in clinical use, to the point that the methicillin-resistant Staphylococcus aureus (MRSA) has been described as a "superbug" because of the threat it poses to health and its relative invulnerability to existing drugs. Response strategies typically include the use of different, stronger antibiotics; however, new strains of MRSA have recently emerged that are resistant even to these drugs. This is an evolutionary arms race, in which bacteria develop strains less susceptible to antibiotics, while medical researchers attempt to develop new antibiotics that can kill them. A similar situation occurs with pesticide resistance in plants and insects. Arms races are not necessarily induced by man; a well-documented example involves the spread of a gene in the butterfly Hypolimnas bolina suppressing male-killing activity by *Wolbachia* bacteria parasites on the island of Samoa, where the spread of the gene is known to have occurred over a period of just five years

Evolution by Means of Natural Selection

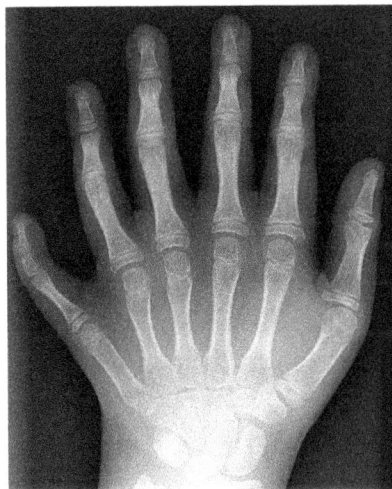

X-ray of the left hand of a ten-year-old boy with polydactyly, caused by a mutant *Hox* gene

A prerequisite for natural selection to result in adaptive evolution, novel traits and spe-ciation is the presence of heritable genetic variation that results in fitness differences. Genetic variation is the result of mutations, genetic recombinations and alterations in the karyotype (the number, shape, size and internal arrangement of the chromosomes). Any of these changes might have an effect that is highly advantageous or highly disad-vantageous, but large effects are rare. In the past, most changes in the genetic material were considered neutral or close to neutral because they occurred in noncoding DNA or resulted in a synonymous substitution. However, many mutations in non-coding DNA have deleterious effects. Although both mutation rates and average fitness effects of mutations are dependent on the organism, a majority of mutations in humans are slightly deleterious.

Some mutations occur in "toolkit" or regulatory genes. Changes in these often have large effects on the phenotype of the individual because they regulate the function of many other genes. Most, but not all, mutations in regulatory genes result in non-vi-able embryos. Some nonlethal regulatory mutations occur in HOX genes in humans, which can result in a cervical rib or polydactyl, an increase in the number of fingers or toes. When such mutations result in a higher fitness, natural selection favours these phenotypes and the novel trait spreads in the population. Established traits are not immutable; traits that have high fitness in one environmental context may be much less fit if environmental conditions change. In the absence of natural selection to preserve such a trait, it becomes more variable and deteriorate over time, possibly resulting in a vestigial manifestation of the trait, also called evolutionary baggage. In many circum-stances, the apparently vestigial structure may retain a limited functionality, or may be co-opted for other advantageous traits in a phenomenon known as preadaptation. A famous example of a vestigial structure, the eye of the blind mole-rat, is believed to retain function in photoperiod perception.

Speciation

Speciation requires a degree of reproductive isolation—that is, a reduction in gene flow. However, it is intrinsic to the concept of a species that hybrids are selected against, opposing the evolution of reproductive isolation, a problem that was recognized by Darwin. The problem does not occur in allopatric speciation with geographically separated populations, which can diverge with different sets of mutations. E. B. Poulton realized in 1903 that reproductive isolation could evolve through divergence, if each lineage acquired a different, incompatible allele of the same gene. Selection against the heterozygote would then directly create reproductive isolation, leading to the Bateson–Dobzhansky–Muller model, further elaborated by H. Allen Orr and Sergey Gavrilets. With reinforcement, however, natural selection can favor an increase in pre-zygotic isolation, influencing the process of speciation directly.

Genetic Basis

Genotype and Phenotype

Natural selection acts on an organism's phenotype, or physical characteristics. Phenotype is determined by an organism's genetic make-up (genotype) and the environment in which the organism lives. When different organisms in a population possess different versions of a gene for a certain trait, each of these versions is known as an allele. It is this genetic variation that underlies differences in phenotype. An example is the ABO blood type antigens in humans, where three alleles govern the phenotype.

Some traits are governed by only a single gene, but most traits are influenced by the interactions of many genes. A variation in one of the many genes that contributes to a trait may have only a small effect on the phenotype; together, these genes can produce a continuum of possible phenotypic values.

Directionality of Selection

When some component of a trait is heritable, selection alters the frequencies of the different alleles, or variants of the gene that produces the variants of the trait. Selection can be divided into three classes, on the basis of its effect on allele frequencies: directional, stabilizing, and purifying selection. Directional selection occurs when an allele has a greater fitness than others, so that it increases in frequency, gaining an increasing share in the population. This process can continue until the allele is fixed and the entire population shares the fitter phenotype. Far more common is stabilizing selection, which lowers the frequency of alleles that have a deleterious effect on the phenotype – that is, produce organisms of lower fitness. This process can continue until the allele is eliminated from the population. Purifying selection conserves functional genetic features, such as protein-coding genes or regulatory sequences, over time by selective pressure against deleterious variants.

Some forms of balancing selection do not result in fixation, but maintain an allele at intermediate frequencies in a population. This can occur in diploid species (with pairs of chromosomes) when heterozygous individuals (with just one copy of the allele) have a higher fitness than homozygous individuals (with two copies). This is called heterozygote advantage or over-dominance, of which the best-known example is the resistance to malaria in humans heterozygous for sickle-cell anaemia. Maintenance of allelic variation can also occur through disruptive or diversifying selection, which favours genotypes that depart from the average in either direction (that is, the opposite of over-dominance), and can result in a bimodal distribution of trait values. Finally, balancing selection can occur through frequency-dependent selection, where the fitness of one particular phenotype depends on the distribution of other phenotypes in the population. The principles of game theory have been applied to understand the fitness distributions in these situations, particularly in the study of kin selection and the evolution of reciprocal altruism.

Selection, Genetic Variation, and Drift

A portion of all genetic variation is functionally neutral, producing no phenotypic effect or significant difference in fitness. Motoo Kimura's neutral theory of molecular evolution by genetic drift proposes that this variation accounts for a large fraction of observed genetic diversity. Neutral events can radically reduce genetic variation through population bottlenecks. which among other things can cause the founder effect in initially small new populations. When genetic variation does not result in differences in fitness, selection cannot directly affect the frequency of such variation. As a result, the genetic variation at those sites is higher than at sites where variation does influence fitness. However, after a period with no new mutations, the genetic variation at these sites is eliminated due to genetic drift. Natural selection reduces genetic variation by eliminating maladapted individuals, and consequently the mutations that caused the maladaptation. At the same time, new mutations occur, resulting in a mutation–selection balance. The exact outcome of the two processes depends both on the rate at which new mutations occur and on the strength of the natural selection, which is a function of how unfavorable the mutation proves to be.

Genetic linkage occurs when the loci of two alleles are in close proximity on a chromosome. During the formation of gametes, recombination reshuffles the alleles. The chance that such a reshuffle occurs between two alleles is inversely related to the distance between them. Selective sweeps occur when an allele becomes more common in a population as a result of positive selection. As the prevalence of one allele increases, closely linked alleles can also become more common by "genetic hitchhiking", whether they are neutral or even slightly deleterious. A strong selective sweep results in a region of the genome where the positively selected haplotype (the allele and its neighbours) are in essence the only ones that exist in the population. Selective sweeps can be detected by measuring linkage disequilibrium, or whether a given haplotype is overrepresented in the population. Since a selective sweep also results in selection of neighbouring alleles, the presence of a block of strong linkage disequilibrium might indicate a 'recent' selective sweep near the center of the block.

Background selection is the opposite of a selective sweep. If a specific site experiences strong and persistent purifying selection, linked variation tends to be weeded out along with it, producing a region in the genome of low overall variability. Because background selection is a result of deleterious new mutations, which can occur randomly in any haplotype, it does not produce clear blocks of linkage disequilibrium, although with low recombination it can still lead to slightly negative linkage disequilibrium overall.

Impact

Darwin's ideas, along with those of Adam Smith and Karl Marx, had a profound influence on 19th century thought, including his radical claim that "elaborately constructed forms, so different from each other, and dependent on each other in so complex a manner" evolved from the simplest forms of life by a few simple principles. This inspired some of Darwin's most ardent supporters—and provoked the strongest opposition. Natural selection had the power, according to Stephen Jay Gould, to "dethrone some of the deepest and most traditional comforts of Western thought", such as the belief that humans have a special place in the world.

In the words of the philosopher Daniel Dennett, "Darwin's dangerous idea" of evolution by natural selection is a "universal acid," which cannot be kept restricted to any vessel or container, as it soon leaks out, working its way into ever-wider surroundings. Thus, in the last decades, the concept of natural selection has spread from evolutionary biology to other disciplines, including evolutionary computation, quantum Darwinism, evolutionary economics, evolutionary epistemology, evolutionary psychology, and cosmological natural selection. This unlimited applicability has been called universal Darwinism.

Origin of Life

How life originated from inorganic matter remains an unresolved problem in biology. One prominent hypothesis is that life first appeared in the form of short self-replicating RNA polymers. On this view, life may have come into existence when RNA chains first experienced the basic conditions, as conceived by Charles Darwin, for natural selection to operate. These conditions are: heritability, variation of type, and competition for limited resources. The fitness of an early RNA replicator would likely have been a function of adaptive capacities that were intrinsic (i.e., determined by the nucleotide sequence) and the availability of resources. The three primary adaptive capacities could logically have been:

(1) The capacity to replicate with moderate fidelity (giving rise to both heritability and variation of type),

(2) The capacity to avoid decay,

(3) The capacity to acquire and process resources.

These capacities would have been determined initially by the folded configurations (including those configurations with ribozyme activity) of the RNA replicators that, in turn, would have been encoded in their individual nucleotide sequences.

Cell and Molecular Biology

In 1881, the embryologist Wilhelm Roux published *Der Kampf der Theile im Organismus* (*The Struggle of Parts in the Organism*) in which he suggested that the development of an organism results from a Darwinian competition between the parts of the embryo, occurring at all levels, from molecules to organs. In recent years, a modern version of this theory has been proposed by Jean-Jacques Kupiec. According to this cellular Darwinism, random variation at the molecular level generates diversity in cell types whereas cell interactions impose a characteristic order on the developing embryo.

Social and Psychological Theory

The social implications of the theory of evolution by natural selection also became the source of continuing controversy. Friedrich Engels, a German political philosopher and co-originator of the ideology of communism, wrote in 1872 that "Darwin did not know what a bitter satire he wrote on mankind, and especially on his countrymen, when he showed that free competition, the struggle for existence, which the economists celebrate as the highest historical achievement, is the normal state of the *animal kingdom*." Herbert Spencer and the eugenics advocate Francis Galton's interpretation of natural selection as necessarily progressive, leading to supposed advances in intelligence and civilization, became a justification for colonialism, eugenics, and social Darwinism. For example, in 1940, Konrad Lorenz, in writings that he subsequently disowned, used the theory as a justification for policies of the Nazi state. He wrote "selection for toughness, heroism, and social utility must be accomplished by some human institution, if mankind, in default of selective factors, is not to be ruined by domestication-induced degeneracy. The racial idea as the basis of our state has already accomplished much in this respect." Others have developed ideas that human societies and culture evolve by mechanisms analogous to those that apply to evolution of species.

More recently work among anthropologists and a psychologist has led to the development of sociobiology and later of evolutionary psychology, a field that attempts to explain features of human psychology in terms of adaptation to the ancestral environment. The most prominent example of evolutionary psychology, notably advanced in the early work of Noam Chomsky and later by Steven Pinker, is the hypothesis that the human brain has adapted to acquire the grammatical rules of natural language. Other aspects of human behavior and social structures, from specific cultural norms such as incest avoidance to broader patterns such as gender roles, have been hypothesized to have similar origins as adaptations to the early environment in which modern humans evolved. By analogy to the action of natural selection on genes, the concept of memes—"units of cultural transmission," or culture's equivalents of genes undergoing selection and recombination—has arisen, first described in this form by Richard Dawkins in 1976 and subsequently expanded upon by philosophers such as Daniel Dennettas explanations for complex cultural activities, including human consciousness.

Information and Systems Theory

In 1922, Alfred J. Lotka proposed that natural selection might be understood as a physical principle that could be described in terms of the use of energy by a system, a concept later developed by Howard T. Odum as the maximum power principle in thermodynamics, whereby evolutionary systems with selective advantage maximise the rate of useful energy transformation.

The principles of natural selection have inspired a variety of computational techniques, such as "soft" artificial life, that simulate selective processes and can be highly efficient in 'adapting' entities to an environment defined by a specified fitness function. For example, a class of heuristic optimisation algorithms known as genetic algorithms, pioneered by John Henry Holland in the 1970s and expanded upon by David E. Goldberg, identify optimal solutions by simulated reproduction and mutation of a population of solutions defined by an initial probability distribution. Such algorithms are particularly useful when applied to problems whose energy landscape is very rough or has many local minima.

In Fiction

Darwinian evolution by natural selection is pervasive in literature, whether taken optimistically in terms of how humanity may evolve towards perfection, or pessimistically in terms of the dire consequences of the interaction of human nature and the struggle for survival. Among major responses is Samuel Butler's 1872 pessimistic *Erewhon* ("nowhere", written backwards). In 1893 H. G. Wells imagined "The Man of the Year Million", transformed by natural selection into a being with a huge head and eyes, and shrunken body.

Heritable Traits

A heritable quantitative trait is a measurable phenotype that depends on the cumulative actions of many genes, and the environment. These traits can vary among individuals, over a range, to produce a continuous distribution of phenotypes.

Examples of Heritable Traits

Earlobe Attachment

If earlobes hang free, they are detached. If they connect directly to the sides of the head, they are attached. Earlobe attachment is a continuous trait: while most earlobes can be neatly categorized as attached or unattached, some are in-between.

Although some sources say that this trait is controlled by a single gene, with unattached earlobes being dominant over attached earlobes, no published studies support this view. Earlobe attachment and shape are inherited, but it is likely that many genes contribute to this trait. As such, its pattern of inheritance is difficult to predict.

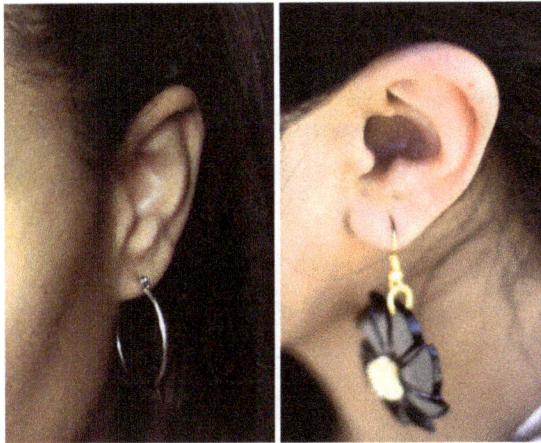

Attached earlobe   Unattached earlobe

Tongue Rolling

Some people can curl up the sides of their tongue to form a tube shape. In 1940, Alfred Sturtevant observed that about 70% of people of European ancestry could roll their tongues and the remaining 30% could not.

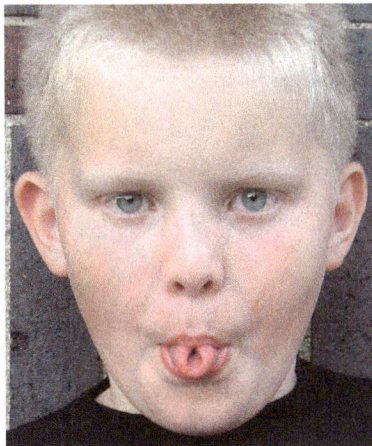

Many sources state that tongue rolling is controlled by a single gene. However, as Sturtevant observed, people can learn to roll their tongue as they get older, suggesting that environmental factors—not just genes—influence the trait. Consistent with this view, just 70% of identical twins share the trait (if tongue rolling were influenced only by genes, then 100% of identical traits would share the trait).

Dimples

Dimples are small, natural indentations on the cheeks. They can appear on one or both sides, and they often change with age. Some people are born with dimples that disappear when they're adults; others develop dimples later in childhood.

Dimples are highly heritable, meaning that people who have dimples tend to have children with dimples—but not always. Because their inheritance isn't completely predictable, dimples are considered an "irregular" dominant trait. Having dimples is probably controlled mainly by one gene but also influenced by other genes.

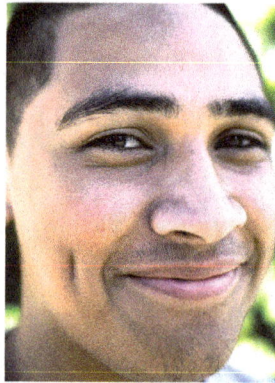

Handedness

Handedness describes our preference for using either our left or right hand for activities such as writing and throwing a ball. Overall, about 10% of people are left-handed, but the number varies among cultures from 0.5% to 24%.

Some have reported that handedness is controlled by just one or two genes, but this is not the case. Multiple studies present evidence that handedness is controlled by many genes—at least 30 and as many as 100—each with a small effect; many are linked to brain development. Environment also plays an important role: some cultures actively discourage left-handedness.

Freckles

Freckles are small, concentrated spots of a skin pigment called melanin. Most fair-skinned, red-haired people have them.

Freckles are controlled primarily by the MC1R gene. Freckles show a dominant inheritance pattern: parents who have freckles tend to have children with freckles.

Variations, also called alleles, of MC1R control freckle number. Other genes and the environment influence freckle size, color, and pattern. For example, sun exposure can temporarily cause more freckles to appear.

Curly Hair

Round hair follicles make straight hair, flattened or c-shaped hair follicles make curly hair, and oval hair follicles make wavy hair. Hair texture is a continuous trait, meaning that hair can be straight or curly or anywhere in between.

Curly hair is influenced by genes much more than by the environment. While curly hair runs in families—people with curly hair tend to have children with curly hair—its inheritance patterns are often unpredictable.

Multiple genes control hair texture, and different variations in these genes are found in different populations. For instance, curly hair is common in African populations, rare in Asian populations, and in-between in Europeans. Straight hair in Asians is mostly caused by variations in two genes—different genes from the ones that influence hair texture in Europeans. And different genetic variations make hair curly in African and European populations.

Hand Clasping

Without thinking about it, fold your hands together by interlocking your fingers. Which thumb is on top—your left or your right?

One study found that 55% of people place their left thumb on top, 45% place their right thumb on top, and 1% have no preference. A study of identical twins concluded that hand clasping has a strong genetic basis (most twins share the trait), but it doesn't fit a predictable inheritance pattern. It is likely affected by multiple genes as well as environmental factors.

Red/Green Colorblindness

Red-green colorblindness is caused by a single gene located on the X-chromosome. This gene codes for a protein in the eye that detects certain colors of light. When this gene is defective, the eye cannot differentiate between red and green.

You need at least one working copy of the gene to be able to see red and green. Since boys have just one X-chromosome, which they receive from their mother, inheriting one defective copy of the gene will render them colorblind. Girls have two X-chromosomes; to be colorblind they must inherit two defective copies, one from each parent. Consequently, red-green colorblindness is much more frequent in boys (1 in 12) than in girls (1 in 250).

Red-green color blindness follows a very predictable recessive, sex-linked inheritance pattern. A woman with one defective copy of the gene and one functional copy, even though she is not colorblind herself, is known as a "carrier." She has a 50% chance of passing the defective copy to each of her children. Half of her sons will be colorblind, and half of her daughters will be carriers.

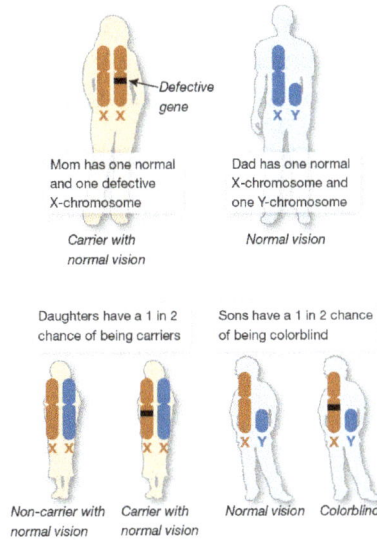

Hairline Shape

If your hairline forms a point at the center of the forehead, you have a widow's peak. If not, you have a straight hairline. While some sources say that widow's peak is a dominant trait controlled by one gene, no scientific study supports this claim. Complicating the question of heritability is the fact that the trait is continuous: some people have just a slight suggestion of a peak.

Widow's peak is likely controlled by genes rather than the environment. But while hairline shape tends to run in families, its pattern of inheritance is usually unpredictable, suggesting that multiple genes are involved.

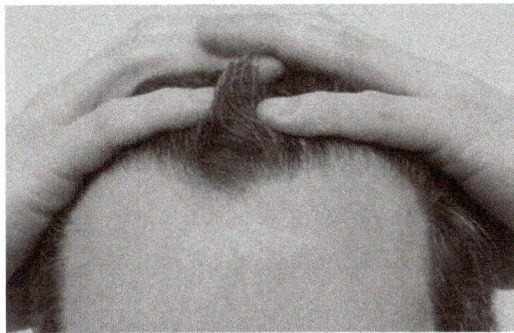

PTC Tasting

To about 75% of us, the chemical PTC (phenylthiocarbamide) tastes very bitter. For the other 25%, it is tasteless. The ability to taste PTC is controlled mainly by a single gene that codes for a bitter-taste receptor on the tongue. Different variations, or alleles, of this gene control whether PTC tastes bitter or not.

PTC tasting follows a very predictable pattern of inheritance. Tasting is dominant, meaning that if you have at least one copy of the tasting version of the gene, you can taste PTC. Non-tasters have two copies of the non-tasting allele.

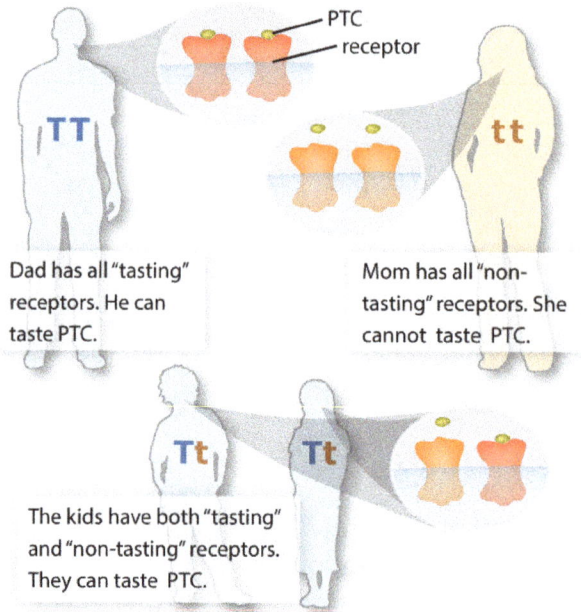

PTC receptor

TT

tt

Dad has all "tasting" receptors. He can taste PTC.

Mom has all "non-tasting" receptors. She cannot taste PTC.

Tt   Tt

The kids have both "tasting" and "non-tasting" receptors. They can taste PTC.

Acquired Traits

Jean-Baptiste Lamarck

Inheritance of acquired characteristics or inheritance of acquired characters is the once widely accepted idea that physiological modifications acquired by an organism can be inherited by the offspring. Acquired characteristics (or characters) are those changes in the structure or function of an organism that are the result of use, disuse, environmental influences, disease, mutilation, and so forth, such as a muscle that is enlarged through use or mice that have their tails cut off. The theory of the inheritance of acquired characteristics, or "soft inheritance," holds that an organism experiencing such a modification can transmit such a character to its offspring.

This theory is commonly equated with the evolutionary theory of French naturalist Jean-Baptiste Lamarck known as Lamarckism. While Lamarck is sometimes viewed as the founder of the concept, in reality this theory was proposed in ancient times by Hippocrates and Aristotle, and Comte de Buffon, before Lamarck, proposed ideas about evolution involving the concept. Even Charles Darwin, after Lamarck, discussed positively this view in his seminal work, *Origin of Species*.

While the theory of the inheritance of acquired characteristics was enormously popular during the early nineteenth century as an explanation for the complexity observed in living systems, after publication of Charles Darwin's theory of natural selection, the importance of individual efforts in the generation of adaptation was considerably diminished. Later, Mendelian genetics supplanted the notion of inheritance of acquired traits, eventually leading to the development of the modern evolutionary synthesis, and the general abandonment of the theory of inheritance of acquired characteristics in biology, although there are proponents for its working on the microbial level or in epigenetic inheritance.

However, in a wider context, the theory of inheritance of acquired characteristics does remain a useful concept when examining the evolution of cultures and ideas and is generally held in terms of some religious views, such as karma and inheritance of sin. In the case of religion, many traditions accept the view that there is an inheritance of acquired spiritual traits—that actions taken in one's life can be passed down in the form of spiritual merit or demerit to one's lineage.

Lamarckism and the inheritance of acquired characteristics

Lamarckism or Lamarckian evolution is a theory put forward by Lamarck based on the heritability of acquired characteristics. Lamarck proposed that individual efforts during the lifetime of the organisms were the main mechanism driving species to adaptation, as they supposedly would acquire adaptive changes and pass them on to offspring.

The identification of "Lamarckism" with the inheritance of acquired characteristics is regarded by some as an artifact of the subsequent history of evolutionary thought, repeated in textbooks without analysis. Stephen Jay Gould wrote that late nineteenth century evolutionists "re-read Lamarck, cast aside the guts of it and elevated one aspect of the mechanics—inheritance of acquired characters—to a central focus it never had for Lamarck himself". He argued that "the restriction of "Lamarckism" to this relatively small and non-distinctive corner of Lamarck's thought must be labeled as more than a misnomer, and truly a discredit to the memory of a man and his much more comprehensive system". Gould advocated defining "Lamarckism" more broadly, in line with Lamarck's overall evolutionary theory.

Lamarck's Theory

The evolution of giraffe necks is often used as the example in explanations of Lamarckism.

Between 1794 and 1796 Erasmus Darwin, grandfather of Charles Darwin, wrote *Zoönomia* suggesting "that all warm-blooded animals have arisen from one living filament," and "with the power of acquiring new parts" in response to stimuli, with each round of "improvements" being inherited by successive generations.

Subsequently Lamarck proposed in his *Philosophie Zoologique* of 1809 the theory that characteristics that were "needed" were acquired (or diminished) during the lifetime of an organism were then passed on to the offspring. He saw this resulting in the development of species in a progressive chain of development towards higher forms.

Lamarck based his theory on two observations, in his day considered to be generally true:

1. Individuals lose characteristics they do not require (or use) and develop characteristics that are useful.

2. Individuals inherit the traits of their ancestors.

With this in mind, Lamarck developed two laws:

1. Law of use and disuse: "In every animal which has not passed the limit of its development, a more frequent and continuous use of any organ gradually strengthens, develops and enlarges that organ, and gives it a power proportional to the length of time it has been so used; while the permanent disuse of any organ imperceptibly weakens and deteriorates it, and progressively diminishes its functional capacity, until it finally disappears."

2. Inheritance of acquired traits: "All the acquisitions or losses wrought by nature on individuals, through the influence of the environment in which their race

has long been placed, and hence through the influence of the predominant use or permanent disuse of any organ; all these are preserved by reproduction to the new individuals which arise, provided that the acquired modifications are common to both sexes, or at least to the individuals which produce the young."

Examples of Lamarckism would include:

- Giraffes stretching their necks to reach leaves high in trees strengthen and gradually lengthen their necks. These giraffes have offspring with slightly longer necks (also known as "soft inheritance").

- A blacksmith, through his work, strengthens the muscles in his arms. His sons will have similar muscular development when they mature.

In essence, a change in the environment brings about change in "needs" *(besoins)*, resulting in change in behavior, bringing change in organ usage and development, bringing change in form over time—and thus the gradual transmutation of the species. While such a theory might explain the observed diversity of species and the first law is generally true, the main argument against Lamarckism is that experiments simply do not support the second law—purely "acquired traits" do not appear in any meaningful sense to be inherited. For example, a human child must learn how to catch a ball even though his or her parents learned the same feat when they were children.

The argument that instinct in animals is evidence for hereditary knowledge is generally regarded within science as false. Such behaviors are more probably passed on through a mechanism called the Baldwin effect. Lamarck's theories gained initial acceptance because the mechanisms of inheritance were not elucidated until later in the nineteenth century, after Lamarck's death.

Several historians have argued that Lamarck's name is linked somewhat unfairly to the theory that has come to bear his name, and that Lamarck deserves credit for being an influential early proponent of the *concept* of biological evolution, far more than for the *mechanism* of evolution, in which he simply followed the accepted wisdom of his time. Lamarck died 30 years before the first publication of Charles Darwin's *Origin of Species*. As science historian Stephen Jay Gould has noted, if Lamarck had been aware of Darwin's proposed mechanism of natural selection, there is no reason to assume he would not have accepted it as a more likely alternative to his "own" mechanism. Note also that Darwin, like Lamarck, lacked a plausible alternative mechanism of inheritance—the particulate nature of inheritance was only to be observed by Gregor Mendel somewhat later, published in 1866. Its importance, although Darwin cited Mendel's paper, was not recognized until the modern evolutionary synthesis in the early 1900s. An important point in its favor at the time was that Lamarck's theory contained a mechanism describing how variation is maintained, which Darwin's own theory lacked.

Proponents

Lamarck founded a school of French *Transformationism* which included Étienne Geoffroy Saint-Hilaire, and which corresponded with a radical British school of comparative anatomy based at the University of Edinburgh, which included the surgeon Robert Knox and the anatomist Robert Edmund Grant. Professor Robert Jameson wrote an anonymous paper in 1826 praising "Mr. Lamarck" for explaining how the higher animals had "evolved" from the "simplest worms"—this was the first use of the word "evolved" in a modern sense. As a young student Charles Darwin was tutored by Grant, and worked with him on marine creatures.

The *Vestiges of the Natural History of Creation,* authored by Robert Chambers and published anonymously in England in 1844, proposed a theory modeled after Lamarckism, causing political controversy for its radicalism and unorthodoxy, but exciting popular interest and paving the way for Darwin.

Darwin's *Origin of Species* proposed natural selection as the main mechanism for development of species, but did not rule out a variant of Lamarckism as a supplementary mechanism. Darwin called his Lamarckian hypothesis Pangenesis, and explained it in the final chapter of his book *Variation in Plants and Animals under Domestication,* after describing numerous examples to demonstrate what he considered to be the inheritance of acquired characteristics. Pangenesis, which he emphasized was a hypothesis, was based on the idea that somatic cells would, in response to environmental stimulation (use and disuse), throw off 'gemmules' which traveled around the body (though not in necessarily in the bloodstream). These pangenes were microscopic particles that supposedly contained information about the characteristics of their parent cell, and Darwin believed that they eventually accumulated in the germ cells where they could pass on to the next generation the newly acquired characteristics of the parents.

Darwin's half-cousin, Francis Galton carried out experiments on rabbits, with Darwin's cooperation, in which he transfused the blood of one variety of rabbit into another variety in the expectation that its offspring would show some characteristics of the first. They did not, and Galton declared that he had disproved Darwin's hypothesis of Pangenesis, but Darwin objected, in a letter to "Nature" that he had done nothing of the sort, since he had never mentioned blood in his writings. He pointed out that he regarded pangenesis as occurring in Protozoa and plants, which have no blood (Darwin 1871). With the development of the modern synthesis of the theory of evolution and a lack of evidence for either a mechanism or even the heritability acquired characteristics, Lamarckism largely fell from favor.

In the 1920s, experiments by Paul Kammerer on amphibians, particularly the midwife toad, appeared to find evidence supporting Lamarckism, but were discredited as having been falsified. In *The Case of the Midwife Toad,* Arthur Koestler surmised that the specimens had been faked by a Nazi sympathizer to discredit Kammerer for his political views.

A form of "Lamarckism" was revived in the Soviet Union of the 1930s when Trofim Lysenko promoted Lysenkoism which suited the ideological opposition of Joseph Stalin to Genetics. This ideologically driven research influenced Soviet agricultural policy which in turn was later blamed for crop failures.

Since 1988 certain scientists have produced work proposing that Lamarckism could apply to single celled organisms. The discredited belief that Lamarckism holds for higher order animals is still clung to in certain branches of new-age pseudoscience under the term racial memory.

Steele produced some indirect evidence for somatic transfer of antibody genes into sex cells via reverse transcription. Homologous DNA sequences from VDJ regions of parent mice were found in germ cells and then their offspring.

Neo-Lamarckism

Unlike neo-Darwinism, the term neo-Lamarckism refers more to a loose grouping of largely heterodoxical theories and mechanisms that emerged after Lamarck's time, than to any coherent body of theoretical work.

In the 1920s, Harvard University researcher William McDougall studied the abilities of rats to correctly solve mazes. He claimed that offspring of rats that had learned the maze were able to run it faster. The first rats would get it wrong an average of 165 times before being able to run it perfectly each time, but after a few generations it was down to 20. McDougall attributed this to some sort of Lamarckian evolutionary process.

At around the same time, Russian physiologist Ivan Pavlov, who was also a Lamarckist, claimed to have observed a similar phenomenon in animals being subject to conditioned reflex experiments. He claimed that with each generation, the animals became easier to condition.

Neither McDougall nor Pavlov suggested a mechanism to explain their observations.

Soma to Germ Line Feedback

In the 1970s, the immunologist Ted Steele, formerly of the University of Wollongong, and colleagues, proposed a neo-Lamarckian mechanism to try and explain why homologous DNA sequences from the VDJ gene regions of parent mice were found in their germ cells and seemed to persist in the offspring for a few generations. The mechanism involved the somatic selection and clonal amplification of newly acquired antibody gene sequences that were generated via somatic hyper-mutation in B-cells. The mRNA products of these somatically novel genes were captured by retroviruses endogenous to the B-cells and were then transported through the blood stream where they could breach the soma-germ barrier and retrofect (reverse transcribe) the newly acquired genes into the cells of the germ line. Although Steele was advocating this theory for the better part of two decades, little more than indirect evidence was ever acquired to support it. An interesting attribute of this idea is that it strongly resembles Darwin's own theory of pangenesis, except in the soma to germ line feedback theory; pangenes are replaced with realistic retroviruses.

Epigenetic Inheritance

Forms of 'soft' or epigenetic inheritance within organisms have been suggested as neo-Lamarckian in nature by such scientists as Eva Jablonka and Marion J. Lamb. In addition to "hard" or genetic inheritance, involving the duplication of genetic material and its segregation during meiosis, there are other hereditary elements that pass into the germ cells also. These include things like methylation patterns in DNA and chromatin marks, both of which regulate the activity of genes. These are considered "Lamarckian" in the sense that they are responsive to environmental stimuli and can differentially effect gene expression adaptively, with phenotypic results that can persist for many generations in certain organisms. Although the reality of epigenetic inheritance is not doubted (as countless experiments have validated it) its significance to the evolutionary process is however uncertain. Most neo-Darwinians consider epigenetic inheritance mechanisms to be little more than a specialized form of phenotypic plasticity, with no potential to introduce evolutionary novelty into a species lineage.

Lamarckism and Single-celled Organisms

While Lamarckism has been discredited as an evolutionary influence for larger life forms, some scientists controversially argue that it can be observed among microorganisms. Whether such mutations are directed or not also remains a point of contention.

In 1988, John Cairns at the Radcliffe Infirmary in Oxford, England, and a group of other scientists renewed the Lamarckian controversy (which by then had been a dead debate for many years). The group took a mutated strain of E. coli that was unable to consume the sugar lactose and placed it in an environment where lactose was the only food source. They observed over time that mutations occurred within the colony at a rate that suggested the bacteria were overcoming their handicap by altering their own genes. Cairns, among others, dubbed the process adaptive mutagenesis.

If bacteria that had overcome their own inability to consume lactose passed on this "learned" trait to future generations, it could be argued as a form of Lamarckism; though Cairns later chose to distance himself from such a position. More typically, it might be viewed as a form of ontogenic evolution.

There has been some research into Lamarckism and prions. A group of researchers, for example, discovered that in yeast cells containing a specific prion protein Sup35, the yeast were able to gain new genetic material, some of which gave them new abilities such as resistance to a particular herbicide. When the researchers mated the yeast cells with cells not containing the prion, the trait reappeared in some of the resulting offspring, indicating that some information indeed was passed down, though whether or not the information is genetic is debatable: trace prion amounts in the cells may be passed to their offspring, giving the appearance of a new genetic trait where there is none.

Finally, there is growing evidence that cells can activate low-fidelity DNA polymerases in times of stress to induce mutations. While this does not directly confer advantage to the organism on the organismal level, it makes sense at the gene-evolution level. While the acquisition of new genetic traits is random, and selection remains Darwinian, the active process of identifying the necessity to mutate is considered to be Lamarckian.

Sexual Selection

Sexual selection is a "special case" of natural selection. Sexual selection acts on an organism's ability to obtain (often by any means necessary) or successfully copulate with a mate.

Selection makes many organisms go to extreme lengths for sex: peacocks (top left) maintain elaborate tails, elephant seals (top right) fight over territories, fruit flies perform dances, and some species deliver persuasive gifts. After all, what female Mormon cricket (bottom right) could resist the gift of a juicy sperm-packet? Going to even more extreme lengths, the male redback spider (bottom left) literally flings itself into the jaws of death in order to mate successfully.

Sexual selection is often powerful enough to produce features that are harmful to the individual's survival. For example, extravagant and colorful tail feathers or fins are likely to attract predators as well as interested members of the opposite sex.

Charles Darwin proposed that all living species were derived from common ancestors. The primary mechanism he proposed to explain this fact was natural selection: that is, that organisms better adapted to their environment would benefit from higher rates of survival than those less well equipped to do so. However he noted that there were many examples of elaborate, and apparently non-adaptive, sexual traits that would clearly not aid in the survival of their bearers. He suggested that such traits might evolve if they are sexually selected, that is if they increase the individual's reproductive success, even at the expense of their survival.

Darwin noted that sexual selection depends on the struggle between males to access females. He recognized two mechanisms of sexual selection: intrasexual selection, or competition between members of the same sex (usually males) for access to mates, and intersexual selection, where members of one sex (usually females) choose members of the opposite sex. The idea of cumbersome traits evolving to aid males in competition during aggressive encounters was readily accepted by scientists shortly after Darwin's publication. However, the idea of female mate choice was received with ridicule, and was not seriously reconsidered until nearly 80 years later. In the 40 years since, there has been much progress in our understanding of how sexual selection operates.

Sex that is under Stronger Selection

Sex roles are defined by differences in gametes: females produce relatively few, highly nutritious (usually non-motile) gametes, whereas males produce comparatively abundant, smaller, motile gametes. Because only a single gamete of each type is required to produce an offspring, there will be an excess of male gametes that will not fertilize any eggs. This asymmetry leads to Bateman's principle, whereby female reproduction is primarily limited by their access to resources to nourish and produce these large gametes, whereas male reproduction is mainly limited by access to females. Therefore males typically compete among themselves for access to females, whereas females tend to be choosy and mate only with preferred males.

In sexually reproducing species, every offspring has one father and one mother, so the average reproductive success is equal for both males and females. A successful male can potentially sire many offspring. If a male gains a disproportionate share of reproduction, he will take away reproductive opportunities from other males, leading to a high reproductive variance among males. A successful female, on the other hand, will not take away reproductive opportunities from other females, leading to a smaller variance in reproductive success. The higher the reproductive variance, the stronger the effects of sexual selection. Strong sexual selection typically results in sexually dimorphic traits that are exaggerated, or more elaborate, in the sex with highest reproductive variance.

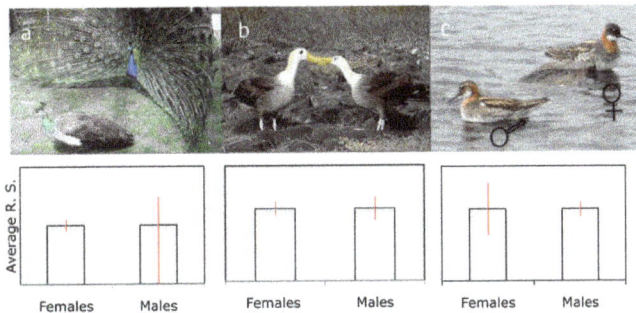

Figure: Variance in reproductive success explains which sex is subject to stronger sexual selection

Males and females in a population have the same average reproductive success (R. S., black bars) but they may differ in the reproductive variance among members of each sex (shown in red). Differences in the selection gradient will result in sexual dimorphism.

(A) When males are subject to stronger sexual selection than females, males will evolve secondary sexual characters that result in marked differences between the sexes. Peacocks do not provide any parental care, and some males are more successful than others who may never reproduce, leading to marked dimorphism.

(B) When males contribute to offspring care, the selection gradient is lower and the sexes will be monomorphic. Many seabirds are monogamous and raise offspring together and the sexes are indistinguishable.

(C) When males provide all the parental care, the selection gradient can be reversed and females may have to compete for access to males, leading to reverse sexual dimorphism. Red-necked phalaropes compete for access to males who provide all the parental care. Females are larger and more aggressive than males.

The Role of Parental Care

Most species provide little or no care to their offspring, but in species where parental care is required, variance in reproductive success will be impacted not only by fertilization success, but also by the contribution of each sex to the care of the offspring. The degree and direction of sexual dimorphism can be explained by the relative selection gradients of each sex. If females provide more parental care than males, the variance in male reproductive success can be expected to be large, since females providing offspring care will not be immediately available for further reproduction and competition for available females will increase among males. The exaggerated tail of the (male) peacock compared to the shorter tail of the (female) peahen, indicates that males are under stronger sexual selection than females. However, in species where biparental care is required to successfully raise offspring, variance in male reproductive success is generally lower, since males that are engaged in providing parental care will not be able to invest as much energy in pursuing additional mating opportunities. This situation often results in the emergence of sexually monomorphic species, in which the male and female look and behave in similar ways. Finally, in rare situations where only males provide parental care, males can become the limiting resource for females. Under these circumstances, the variance in reproductive success may be high for females who then tend to monopolize access to one or more males to care for their offspring. Males may then become choosy about which females they mate with. This often results in reversed sexual dimorphism, such that females evolve more elaborate secondary sexual characters than males.

Mechanism of Sexual Selection Operate

Sexual selection can operate both intra- and inter-sexually, either sequentially or simultaneously. During intrasexual selection, members of the same sex attempt to outcompete rivals, often during direct encounters. Intrasexual selection is typically responsible for the evolution of male armaments such as deer antlers, beetle horns, and large body size that provide individuals with an advantage when fighting off potential competitors. Individuals, who are better able to exclude competitors, have a greater chance to acquire mates and father offspring. For example, dominant male red deer monopolize a group of females (also known as harem) by constantly fighting off competitors, and they father most of the offspring produced by the females. By contrast, intersexual selection results from interactions between the sexes, typically involving mate choice. The evolution of elaborate behavioral displays and morphological traits can often be explained as the result of intersexual selection. Usually, females tend to be more choosy, evaluating morphological and behavioral traits from potential mates to determine which will maximize their fitness. Males tend to compete with one another to gain the female's attention. An extreme example of intersexual selection can be found in species where males form leks where multiple males gather to display to females.

Sexual selection episodes can occur before mating takes place (pre-copulatory), or during and after mating (post-copulatory), and they can occur within a sex (intrasexual) and between the sexes (intersexual).

	Prc-cooulato""	Post conulato...,
lntrascxuat SClcct on (competition)	• To monopolize a harem: Elephant seals (Le Boeuf 1974) • To monopolize a sing e fema e: Rhinoceros Beetles (Emlen 2008)	• Greater relative testis size in species with greater risk of female mutiple mating (Calh m and Blrkheod 2007)
Interscxual Selectlon (mate choice)	• Males displaying at leks in Manakins (Prum 1990)	• Female reproductive morphology that prevents eja-culated sperm from contacting eggs directly (Keler and Reeve 1995)
SexualConflict	• Intimidation and sexual harassment (Clutton-Brock and Parker 1995)	• Traumatic insemination in bed bugs (Stutt and Sivajothy 2001)

Table: Some examples of when and how sexual selection operates

Choosing a Mate

Why do females choose between males rather than mate at random, or with the first male they encounter? Females can directly increase their reproductive success by mating with certain, select males and acquiring direct benefits. For example, females can gain increased access to food, protection from harassing males, or help in raising offspring, and avoid being infected with parasites or other diseases by choosing healthy males. However there are instances where females do not appear to gain any direct benefit from males, yet they still discriminate among them. Under these conditions, females likely gain indirect benefits via their offspring. These indirect benefits are usually genetic rather than resource based. By choosing certain males, their offspring will likely inherit genes that tend to increase their fitness. Males often evolve traits and displays that advertise their ability to provide direct and indirect benefits, and females evolve preferences for these traits. Two major mechanisms to account for female mate choice have been proposed: good genes, and Fisherian arbitrary processes.

Good Genes

Under the 'good genes' scenario, differences among males provide females with information about the genetic qualities of the different males that can be inherited by the offspring. Under the 'good genes,' just as in the 'direct benefits' models, there is correspondence between the putative roles of natural versus sexual selection, since preferring certain males can result in a female gaining higher viability, fecundity, and reproductive success, for her offspring. Good genes can be those that allow males to carry a 'handicap,' yet survive despite having a cumbersome trait, genes that signal resistance to diseases, or genes that are more compatible with those of the female. Evidence of female choice for good genes remains scarce despite decades of studies of female mate choice in many taxa. This apparent lack of success continues to create debate as to the importance of the good genes model in the field.

Fisherian Arbitrary Choice

Named after R. A. Fisher, who originally proposed it, this model suggests that female preference can evolve for arbitrary traits that do not provide information about the male's quality, and that therefore do no reinforce the effects of natural selection. If females evolve a preference for a particular trait, males bearing that trait will be selected as mates. This assortative mating will establish a genetic correlation between the preference and the trait. The fitness advantage of the arbitrary trait exists only as a result of its covariance with the preference. By choosing a male with a particular trait value, the females gain the indirect benefit of producing offspring that will be more sexually attractive to females that carry the preference. This type of process can result in a runaway positive feedback loop, whereby the trait becomes more exaggerated as selection on the preference increases, but other models have shown that such a feedback loop is only one of many possible evolutionary outcomes of the Fisherian process.

Instances when Sexual Selection Act

Sexual selection can affect reproductive success at multiple reproductive stages. First, it acts during all the processes that lead to acquiring mating opportunities (i.e., excluding competitors, attracting, selecting and retaining mates). Darwin referred exclusively to pre-copulatory sexual selection in his discussions, erroneously assuming that mating would inevitably result in reproductive success. In recent years, evidence that copulatory and post-copulatory events play an important role in determining the outcome of fertilization and reproduction has been increasing. Post-copulatory selection refers to the events that occur during and after mating. Post-copulatory male-male competition is known as sperm competition a term coined by Parker who recognized that when females mate with multiple males, their ejaculates compete inside the female reproductive tract for access to eggs. Sperm competition has resulted in the evolution of morphologically modified sperm that increase the likelihood of fertilization in many taxa. Post-copulatory female choice refers to the ability of females to affect the likelihood that sperm from a particular male fertilizes their eggs, and their decision to invest in offspring based on the identity of the male with whom they mate. Females exert this choice via morphological, chemical and behavioral adaptations. This type of selection is called cryptic choice because it occurs inside the female reproductive tract and cannot be detected from behavioral studies alone.

Conflict between the Sexes

Although both sexes are seeking to optimize their reproductive success, their genetic interests are not aligned, resulting in sexual conflict. Traits that allow a male to increase his reproductive success at the expense of the female will be positively selected if the female mates with multiple males. These traits will be genetically transmitted and spread in the population, despite their negative effects on female reproductive success, if the reproductive success of these males is higher than that of males lacking such traits. Sexual conflict can often result in an evolutionary arms race, whereby the evolution of a trait that imposes harm on one sex will result in evolution of a counter-trait to mitigate the harm on the affected sex, with subsequent escalation in both. Examples of sexual conflict include traumatic insemination in bed bugs, copulatory grasping and anti-grasping structures in waterstriders, and genital coevolution in waterfall.

Stabilizing Selection

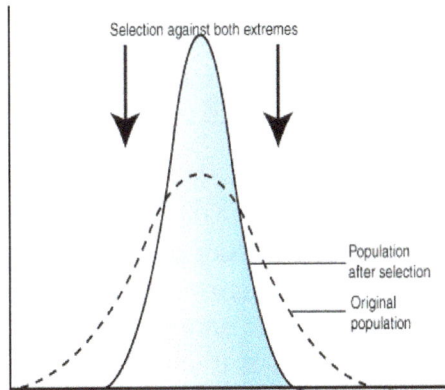

Selection against both extremes

Population after selection

Original population

(a) **Stabilizing selection**

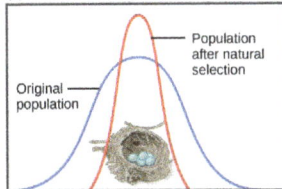

Population after natural selection

Original population

Robins typically lay four eggs, an example of stabilizing selection. Larger clutches may result in malnourished chicks, while smaller clutches may result in no viable offspring.

(b) **Directional selection**

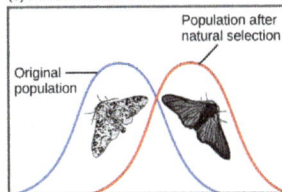

Population after natural selection

Original population

Light-colored peppered moths are better camouflaged against a pristine environment; likewise, dark-colored peppered moths are better camouflaged against a sooty environment. Thus, as the Industrial Revolution progressed in nineteenth-century England, the color of the moth population shifted from light to dark, an example of directional selection.

(c) **Diversifying selection**

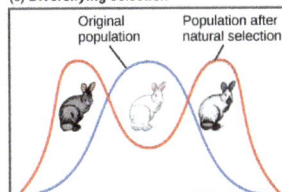

Original population   Population after natural selection

In a hyphothetical population, gray and Himalayan (gray and white) rabbits are better able to blend with a rocky environment than white rabbits, resulting in diversifying selection.

Stabilizing Selection Examples

Stabilizing selection is any selective force or forces which push a population toward the average, or median trait. Stabilizing selection is a descriptive term for what happens to an individual trait when the extremes of the trait are selected against. This increases the frequency of the trait in the population, and the alleles and genes which help form it. Many traits which are common across entire groups of species have been formed through the effects of stabilizing selection. Stabilizing selection can be seen in the image below, comparing the three types of selection.

Robin Eggs

In this case, the number of eggs in a robin nest has been selected for through a stabilizing selection. Robins are apparently not ably to raise more than 4 chicks with much success. This is probably because of the size of the birds and the amount of food that two adults can provide. By comparison, most penguins can only raise one chick at a time, due to the size of the chick and the amount of food it requires. While they stabilized on different numbers, both are forms of stabilizing selection which maximize the fitness of the species in their environment.

As opposed to the other forms of selection, you can clearly see in the stabilizing selection graph that the population of the median trait increases, while the other populations decrease. In this case, 5 eggs is too many and some would die. On the other hand, 3 is too few. Either the eggs are not viable enough to only rely on 3 eggs, or predation and other forces require more than 3 eggs to overcome them and carry on to another generation.

Hypothetical Lemurs

There is a population of multicolored lemurs on Crazy Island. This particular population of lemurs has been observed by scientists, and they have noticed the following changes in the lemur's color.

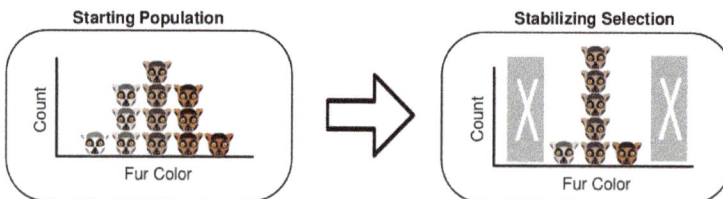

As you can see, the lemurs have obviously undergone a stabilizing selection. The light and dark lemurs have almost disappeared, while the middle brown lemurs have increased. Without further information, it is not clear why this would be the case. It is the job of ecologists and evolutionary biologists to observe the population, noting peculiar aspects of the various forms to understand what may have caused the stabilizing selection. This is no easy question to answer and may have many more than one answer.

In the case if the lemurs, it could be that the darker and lighter lemurs were both easier to spot by predators. If the lemurs only have one predator, this is as easy hypothesis to test. A scientist would simply observe the predator, and see which lemurs it prefers. This would lend evidence to the hypothesis that the stabilizing selection is caused by *predation*. Further evidence could include the amount of lemurs the predators eat, and models showing how that level of predation could produce the coloration seen.

However, it is much more common that a species has multiple selective pressures, and that each pressure acts on various traits in different ways. For example, the lighter color could be suffering from predation, whereas the darker version could be overheating. (Dark colors absorb more solar heat). Likewise, predation could be driving both traits, but not influencing them totally. Female lemurs might have a preference for brown lemurs, due to their increased survival. This would be a form of *sexual selection*, driving a trend of stabilizing selection.

Common Causes of Stabilizing Selection

Stabilizing selection, along with *directional selection* and *disruptive selection*, refer to the direction of individual traits. While stabilizing selection pushed the trait towards the average instead of one or both of the extremes, it can be driven by any form of selection. Some of the most common forms of selection are from predation, resource allocation, coloration of the environment, food type, and a wide variety of other forces.

Many traits we don't talk about regularly have been driven by a variety of causes throughout history. Take the modest insect, for example. All insects have an *exoskeleton*, a miraculous structure made of *chitin* and other structural molecules which form a shield around their organs and allow them to maintain a water balance in the harshest of environments. This shield, while it has been modified in to an almost infinite number of forms, was first selected for out of stabilizing selection. The ancestors to insects did not have this *adaptation*, but once it evolved it was highly favored.

Simply stated, there is no common cause of stabilizing selection, besides the fact that the most average individual is selected for. In that way, like all forms of selection, the cause of stabilizing selection is the increased fitness and reproductive success that the median individuals have. The extreme versions or traits have a disadvantage, in one way or another. This disadvantage, in evolutionary terms, is decreased reproduction. The traits they carry are coded for in part by their DNA, which they can only pass on through reproduction. In stabilizing selection, the increase in the median traits represents their increased success. The other extreme traits are not as successful, possibly causing their owners to die. This increases the resources available to the median animals, further boosting their success. In this way, stabilizing selection is the cause of many traits that entire groups of animals share. These are known as *synapomophies*.

Directional Selection

Directional selection occurs when individuals with traits on one side of the mean in their population survive better or reproduce more than those on the other. It has been demonstrated many times in natural populations, using both observational and experimental approaches. Directional selection does the "heavy lifting" of evolution by tending to move the trait mean toward the optimum for the environment. It results in increased adaptedness of organisms. It is the principle process that Charles Darwin himself envisaged as driving adaptive evolution. Two of Darwin's own examples were:

(1) Faster wolves being more successful at hunting deer,

(2) Flowers that produce more nectar being more successful in attracting pollinating insects.

These both suggest directional rather than other forms ("modes") of selection. Directional selection is the process that comes most easily to mind when thinking about natural selection, and it is the form of selection that has taken place in the best-known examples of evolution (e.g., the peppered moth, antibiotic resistance, finch beaks). However, directional selection does not always result in evolution, because it can be constrained in many ways. If directional selection acts in different directions in different populations or species, because of variation in environmental circumstances, then it is described as *divergent*. This results in populations becoming different, and it can contribute to speciation. Directional selection can also be artificially imposed, and it has commonly been used by animal and plant breeders to improve traits (such as yield) in domesticated organisms, as well as to better understand evolution.

• One extreme trait is favoured

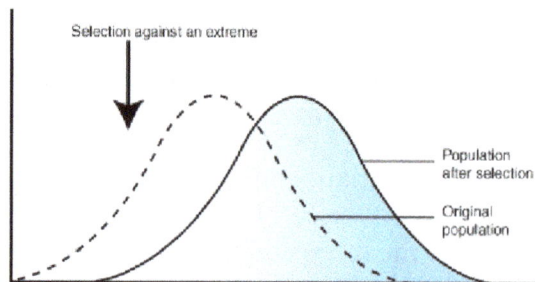

Example: Industrial Melanism.

- In England, before industrialisation, white-winged moth were more in no. than dark-winged moths.

- But after industrialization, dark-winged moths became more in no. than white winged moths.

- This is because during industrialization, the tree trunks covered by white lichen became dark due to deposition of coal & dust particles.

- As a result, white-winged moths can be easily picked up by predators from the dark background & dark-winged moths survived.

Negative Selection

Scientists talk about positive selection when the focus of a particular study is on an increase in rare variants that improve optimal fitness, and they speak of negative selection when the focus is on the removal of harmful variants.

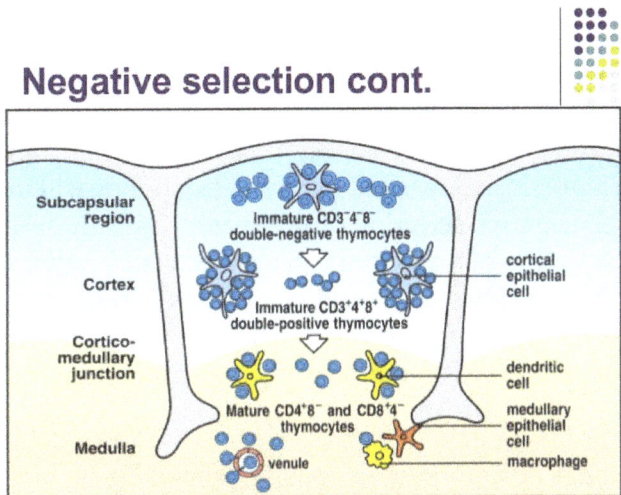

Negative selection cont.

Causes of Negative Selection

Because more DNA changes are harmful than are beneficial, negative selection plays an important role in maintaining the long-term stability of biological structures by removing deleterious mutations. Thus, negative selection is sometimes also called purifying selection or background selection. One key reason why this form of selection is so prevalent is the success of evolution in optimizing biological structures: As soon as a system has been improved, there is the danger of losing that improvement by a deleterious mutation. Purifying selection makes sure that deleterious mutations cannot take over a population and that any improved structures—once fixed in a population—are maintained as long as they are needed. A dramatic example of such maintenance can be found in so-called "living fossils": If a species' ecological niche happens not to change for millions of years, fossil forms of the species can be almost indistinguishable from their present-day descendants.

More short-term negative selection is also widespread, especially due to ecological reasons. Many structures in biology are only conditionally optimal, because they depend upon the details of other structures or circumstances to perform their function. If such other structures are within the same organism, this relationship is termed epistasis. For example, two proteins could interact epistatically in such a way that a deleterious mutation in one protein could be either compensated for or aggravated by a mutation in the other protein. Frequently, ecological circumstances also play a role in determining mutational effects. For instance, if the niche of a species stays the same, some mutations that would be beneficial in other niches will be under negative selection. If the niche changes, however, some traits that were previously under negative selection may suddenly be beneficial and have a greater fitness than the majority of the previously favored genotypes.

If environmental interactions include other rapidly evolving species, then the pressure to change may never stop, and the evolutionary optimum will always remain some steps ahead. Host-parasite interactions are a famous example of this sort of situation. Here, the host immune system evolves to recognize a special structure on the parasite and allow its removal. This in turn induces negative selection on the current form of the parasite while leading to positive selection of variants that cannot be recognized by the host. Furthermore, if such variant parasites exist, they will increase in frequency and in turn induce negative selection of the current variant of the host, which will lead to the positive selection of hosts that can again recognize the parasite, and so on. Negative (and positive) selection in such a system never rests, which is why one hypothesis describing these systems was named after the Red Queen in the book *Alice in Wonderland*, who famously stated that it takes a great amount of running to stay in the same place.

Figure: Long term stabilizing selection can explain \"living fossils.\"

Amber fossils sometimes conserve ancient insect species that are very similar to their modern relatives. Voltinia dramba is the first butterfly to be described from a 20 million-year-old fossil. One of its several close relatives is found in Mexico today.

Effective Strength

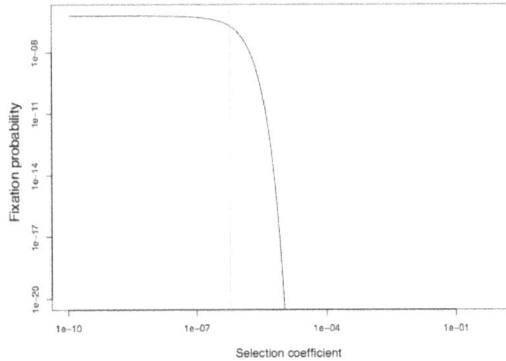

The strength of negative selection

The fixation probability is given for heterozygous, harmful mutations. The vertical line indicates selection coefficients that satisfy Nes = 0.5 and roughly separates the domains where genetic drift (left) and selection (right) dominate the fate of new mutations. The effective population size Ne was estimated from DNA sequence diversity data in the fruitfly *Drosophila miranda*.

There are two ways to measure the strength of negative selection: an absolute one and a relative one. The absolute selection coefficient *s* quantifies the relative fitness difference between the rare variant (mutant) and the most common variant (wild type). The advantage of this measure is its independence from population size. For mutations with large effects, it might even be possible to measure selection coefficients by comparing direct counts of surviving offspring from many individuals with and without a well-specified mutation.

However, most selection coefficients are very small and, thus, selection as a process is often better understood by looking at the relative effective selection coefficient $N_e s$, which is the product of the absolute selection coefficient and the effective population size N_e. A value of 1 ($N_e s = 1$) approximately denotes a threshold: Mutations with $N_e s < 0.25$ are fixed with probabilities that are comparable to those of neutral mutations, even if they are harmful. On the other hand, mutations in which $N_e s > 4$ will very rarely become fixed in a population with reasonable levels of recombination. Moreover, it is unlikely that a harmful mutation with $N_e s$ exceeding a few dozen could have ever been fixed in the entire history of the known universe. Such deleterious mutations can only be fixed if they are closely linked to a very strongly advantageous mutation that will drag them to high frequencies (but, as discussed later, this reduces N_e and, thus, the rule of thumb remains valid). If harmful mutational effects are smaller than roughly $N_e s = 0.25$, then they can pass below the "radar" of natural selection: Here, natural selection can no longer distinguish among beneficial, neutral, and harmful mutations, so it treats them all in the same way. This problem is balanced by the increasing frequency of advantageous back mutations for such sites. To apply this simple rule of thumb correctly, it is important to note that N_e usually has to be measured from DNA sequence diversity data in the wild. N_e is usually much lower than the number of all existing individuals. Also, this rule has limitations when recombination is rare or absent. In this case, the effective strength of selection is reduced, and more deleterious mutations can fix more easily. Figure shows the fixation probabilities of various deleterious mutations under free recombination.

Consequences of Negative Selection

The main consequence of negative selection is the extinction of less-adapted variants. If the best-adapted variant does not change because it is at a stable local optimum, then negative selection will remove all new variants for that optimal trait.

It is important to note that negative selection can also impact molecular diversity. Consider the simple case of a population with genes that are optimally adapted to the existing constant environment. Such a setting is probably realistic for many "housekeeping" genes that ensure the proper working of the basic molecular machinery of life. Almost every mutation that happens in these genes will be deleterious, and, because mutations are the inevitable consequence of the molecular machinery that copies DNA, we can expect a substantial number of such harmful mutations. The deleterious nature of these mutations will result in their quick removal. In any real-life setting, however, an important side effect of such a removal will be the accompanying removal of linked mutations. This has important implications for the study of molecular diversity, as all neutral mutations linked to deleterious mutations will not be observed in the population and the corresponding N_e values will thus be reduced. This reduction in N_e has consequences for adaptive evolution, because the effective strength of selection for positive mutations will be reduced as well, and more advantageous mutations with small effects will be lost by chance.

If negative selection is too weak to remove harmful mutations, then deleterious mutation accumulation will occur, and a gradual decay of genomic integrity will be the result. This can lead to extinction for some species if it continues long enough; however, the resulting widespread existence of deleterious mutations in such a genome will eventually also lead to the occurrence of back mutations, which (among many other factors) can significantly contribute toward maintenance of a reasonable level of integrity in the genome of other species in the long term.

If negative selection is too strong for the whole population, extinction will occur, unless the population is rescued in time. Extinction can occur if the negative selection considered is "hard" selection, which actually reduces the number of surviving offspring that are produced. "Soft" selection (which occurs when the reproductive capacity of an organism is high enough) can also be negative, but it will lead only to competition over who will increase in frequency within the population, effectively without a reduction of the maximal number of offspring that can be produced. Thus, no extinction risk exists with soft selection.

Climate change and other habitat alterations are currently placing many species under such extreme negative selection that these species' survival is threatened. Also, mutagenic substances that are released into the environment by humans lead to a general increase in the frequency of mutations, a vast majority of which are deleterious and further increase the negative selection pressure for many populations. Thus, understanding the causes, extent, and consequences of negative selection can contribute important insights toward securing biodiversity in the long term.

Gene Selection

Genetic selection is the process by which certain traits become more prevalent in a species than other traits. These traits seen in an organism are due to the genes found on their chromosomes. The genes code for the traits that we are able to observe.

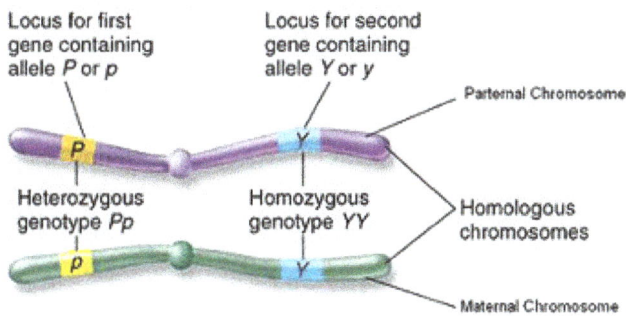

Figure: Alleles for genes are inherited and come in various forms

Genes have more than one version or allele. We inherit one allele for every gene from each of our parents.

Some alleles are seen more frequently in a population because there are factors that select those genes.

In natural selection, natural forces determine the traits seen in an organism. A variation or allele of a trait makes some individuals more suited to survive in the environment. Mating behavior that leads to a sexual preference for a trait is also natural selection.

If you are a mouse and you can blend in with the environment you are likely to live long enough to pass on your genes for coat color to baby mice. Your friend who could not blend in gets eaten and does not pass on the genes for their coat color. Over time, the mice in dark areas will have dark fur and the mice in light areas will have light fur. Natural selection by the environment eliminates the weakest individuals from the gene pool. If the mice did not have a gene for coat color that allowed them to hide from predators, that species of mouse would be eliminated from the environment.

Natural selection also occurs when a species has a preference for certain traits for sexual reasons. Male peafowl (peacocks) have elaborate tail feathers because female peacocks are attracted to really nice tails. This is called sexual selection since traits are being selected for sexual reasons.

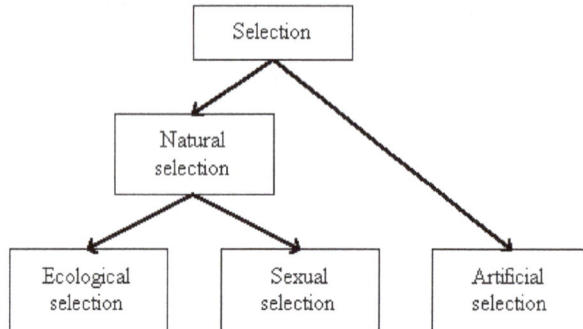

Figure: There are two types of selection: 1) Natural and 2) Artificial.

Gene flow is the movement of genes into or out of a population. Such movement may be due to migration of individual organisms that reproduce in their new populations, or to the movement of gametes (e.g., as a consequence of pollen transfer among plants). In the absence of natural selection and genetic drift, gene flow leads to genetic homogeneity among demes within a metapopulation, such that, for a given locus, allele frequencies will reach equilibrium values equal to the average frequencies across the metapopulation. In contrast, restricted gene flow promotes population divergence via selection and drift, which, if persistent, can lead to speciation.

Gene Selection Theory

Two-toed sloths

In modern evolutionary theory, all evolutionary processes boil down to an organism's genes. Genes are the basic "units of heredity," or the information that is passed along in DNA that tells the cells and molecules how to "build" the organism and how that organism should behave. Genes that are better able to encourage the organism to reproduce, and thus replicate themselves in the organism's offspring, have an advantage over competing genes that are less able. For example, take female sloths: In order to attract a mate, they will scream as loudly as they can, to let potential mates know where they are in the thick jungle. Now, consider two types of genes in female sloths: one gene that allows them to scream extremely loudly, and another that only allows them to scream moderately loudly. In this case, the sloth with the gene that allows her to shout louder will attract more mates—increasing reproductive success—which ensures that her genes are more readily passed on than those of the quieter sloth.

Essentially, genes can boost their own replicative success in two basic ways. First, they can influence the odds for survival and reproduction of the organism they are in (individual reproductive success or fitness—as in the example with the sloths). Second, genes can also influence the organism to help other organisms who also likely contain those genes—known as "genetic relatives"—to survive and reproduce (which is called inclusive fitness). For example, why do human parents tend to help their own kids with the financial burdens of a college education and not the kids next door? Well, having a college education increases one's attractiveness to other mates, which increases one's likelihood for reproducing and passing on genes. And because parents' genes are in their own children (and not the neighborhood children), funding their children's educations increases the likelihood that the parents' genes will be passed on.

Disruptive Selection

Disruptive selection is an evolutionary force that drives a population apart. The disruptive selection will cause organisms with intermediate traits to reproduce less, and will allow those organisms with extreme traits to reproduce more. This causes the alleles for the extreme traits to increase in frequency. Over time, and with enough disruptive selection, a population can be completely divided. When this happens, the two populations can become diverse enough to form separate species. In the most basic sense, disruptive selection can act on a single gene, selecting between the different alleles present in a population. On a much broader level, disruptive selection can affect a variety of traits and drive a population to become reproductively isolated from the original population.

Disruptive selection, also called *diversifying selection*, is based on the *variance* of a trait in a population. A gene with only one allele would have no variance, and selection could not act on differences in the trait created by the gene. Most genes have many different alleles, which create a wide variety of functions. The result of these many alleles acting in a population and the other genes in affect lead to traits that have a distribution of different types, sizes, or patterns. If the trait has an almost infinite variety of different forms, it is *continuous*. If the trait exists in distinct entities, it is *discrete*. A continuous trait would be height, while a discrete trait might be eye color. Either way, most traits have a high level of variance, due to the interactions of various genes and alleles. Disruptive selection acts on the traits in the middle of the spectrum.

Disruptive selection is usually seen in high-density populations. In these populations resources become scarcer, and competition for the resources increases. This *intraspecific competition* can cause differences between organisms to have a more profound effect on each organism's survival. Selective pressures that might not have factored into a low-density population can take effect, and the resulting disruptive selection can drive a population apart. In doing so, the populations are often pushed to different niches, lowering the competition between them. This leads to *sympatric speciation*, or speciation that occurs while populations occupy the same area.

Examples of Disruptive Selection

Finches on Santa Cruz Island

Darwin's finches, or Galapagos finches, are a group of finches that inhabit the long chain of islands known as the Galapagos, famously visited by Charles Darwin. The birds have been rigorously studied, and various patterns of evolution have been seen in different populations on different islands. On Santa Cruz Island, disruptive selection was seen causing speciation in the population of finches that resides there. Due to forces of disruptive selection, intermediate beak sizes have been selected against for generations. The resulting population has almost no medium sized beaks. Beak size is important for more than just gathering food, and it has been found that beak size also changes the mating calls of the various finches. Researchers have found that the populations of birds, once a single population, have genetically diverged and are on the tipping point of being considered separate species.

Disruptive Selection in Plants

Famous biologist John Maynard Smith proposed disruptive selection as a method for plant speciation in the late 1960s. The idea is simple and have been applied to many examples since. Many plant traits, such as the color of pea pods, are controlled by individual genes. In a scenario where disruptive selection is affecting a population of plants, the most intermediate individuals are often the heterozygous individual, or those that contain different types of alleles for a gene. Homozygous individuals, on the other hand, have two of the same alleles for a trait. Whether the allele is functional or not, two of the same will produce a phenotype on the extreme end of the spectrum. These individuals will be protected during the disruptive selection, and reproduce more. Over time, the organisms may differ so much that they become reproductively isolated. Often the intermediates were serving the function of transferring genes between the two populations. Without them, in the presence of disruptive selection, speciation can occur.

Microevolution

Microevolution refers to evolution that occurs at or below the level of species, such as a change in the gene frequency of a population of organisms or the process by which new species are created (speciation). Microevolutionary changes may be due to several processes: mutation, gene flow, genetic drift, and natural selection.

Biologists distinguish between microevolution and macroevolution, the other main class of evolutionary phenomena. Macroevolution refers to evolution that occurs above the level of species, such as the origin of different phyla, the evolution of feathers, the development of vertebrates from invertebrates, and the explosion of new forms of life at the time of the Cambrian explosion.

However, microevolution also has been defined as only including evolutionary change below the level of species, not the process of speciation. When used in this manner, speciation is considered the purview of macroevolution.

Observable instances of evolution are all examples of microevolution; for example, bacterial strains that have become resistant to antibiotics, or color changes in moths over time. Because microevolution can be observed directly, it is widely accepted, unlike macroevolution, which has engendered controversy since the time of Darwin.

Population genetics is the branch of biology that provides the mathematical structure for the study of the process of microevolution.

Overview and Evidences

Evolution can be defined as any heritable change in a population of organisms over time, or, in terms of alleles (alternative forms of genes), as any change in the frequency of alleles within a population. Both small changes, such as a slight increase in the numbers of antibiotic-resistant bacteria in a population of bacteria exposed to an antibiotic, or large changes, such as the development of vertebrates from invertebrates, qualify as evolution.

Microevolution refers to the small heritable changes that occur within a population or species.

Microevolution has been observed in both the laboratory and the field.

Laboratory Evidences of Microevolution

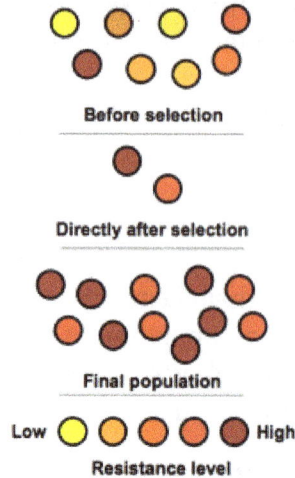

A change in the proportion of antibiotic-resistant bacteria in a population, after exposed to an antibiotic, is an example of microevolution.

In the laboratory, biologists have demonstrated microevolution involving organisms with short lifecycles, such as fruit flies, guppies, and bacteria, which allow testing over many generations.

Endler set up populations of guppies (*Poecilia reticulata*) and their predators in artificial ponds in the laboratory, with the ponds varying in terms of the coarseness of the bottom gravel. Guppies have diverse markings (spots) that are heritable variations and differ from individual to individual. Within 15 generations in this experimental setup, the guppy populations in the ponds had changed according to whether they were exposed to coarse gravel or fine gravel. The end result was that there was a greater proportion of organisms with those markings that allowed the guppies to better blend in with their particular environment, and presumably better avoid being seen and eaten by predators. When predators were removed from the experimental setup, the populations changed such that the spots on the guppies stood out more in their environment, likely to attract mates, in a case of sexual selection.

Likewise, bacteria grown in a Petri dish can be given an antibiotic, such as penicillin, that is just strong enough to destroy most, but not all, of the population. If repeated applications are used after each population returns to normal size, eventually a strain of bacteria with antibiotic resistance may be developed. This more recent population has a different allele frequency than the original population, as a result of selection for those bacteria that have a genetic makeup consistent with antibiotic resistance.

Evidences in the Field

In the field, microevolution has also been demonstrated. Both antibiotic-resistant bacteria and populations of pesticide-resistant insects have been frequently observed in the field. In England, a systematic color change in the peppered moth, *Biston betularia,* has been observed over a 50-year period. While there is some controversy whether this later case can be attributed to natural selection, the evidence of a change in the gene pool over time has been demonstrated. Since the introduction of house sparrows in North America in 1852, they have developed different characteristics in different locations, with larger-bodied populations in the north. This is assumed to be a heritable trait, with selection based on colder weather in the north.

A well-known example of microevolution in the field is the study done by Peter Grant and B. Rosemary Grant on Darwin's finches. They studied two populations of Darwin's finches on a Galapagos island and observed changes in body size and beak traits. For example, after a drought, they recorded that survivors had slightly larger beaks and body size. This is an example of an allele change in populations—microevolution. It is also an apparent example of natural selection, with natural selection defined according to Mayr as: "the process by which in every generation individuals of lower fitness are removed from the population." However, the Grants also found an oscillating effect: when the rains returned, the body and beak sizes of the finches moved in the opposite direction.

Artificial Selection

For thousands of years, humans have artificially manipulated changes within species through artificial selection. By selecting for preferred characteristics in cattle, horses, grains, and so forth, various breeds of animals and varieties of plants have been produced that are different in some respect from their ancestors. This also represents an example of microevolution, in that the changes coming from artificial selection are all within the level of the species.

Hardy-Weinberg Equilibrium

Interestingly, in a nonevolving population, the *allele frequency, genotype frequency,* and *phenotype frequency* remain in *genetic equilibrium.* In other words, the random assortment of genes during sexual reproduction does not alter the genetic makeup of the *gene pool* for that population. This phenomenon was illuminated by the mathematical reasoning of a German physician, Weinberg, and a British mathematician, Hardy, both working independently in 1908. Their combined efforts are now known as the Hardy-Weinberg equilibrium model.

To demonstrate the Hardy-Weinberg equilibrium, assume G and g are the dominant and recessive alleles for a trait where GG = green, gg = yellow, and Gg = orange. In our imaginary population of 1,000 individuals, assume that 600 have the GG genotype, 300 are Gg, and 100 are gg. The allele and genotype frequency for each allele is calculated by dividing the total population into the number for each genotype:

- GG = 600/1,000 =.6

- Gg = 300/1,000 =.3

- gg = 100/1,000 =.1

After the allele frequency has been determined, we can predict the frequency of the allele in the first generation of offspring.

First, determine the total number of alleles possible in the first generation. In this imaginary case, because each organism has 2 alleles and there are 1,000 organisms, the number of possible alleles in the first generation of offspring is:

- 2 × 1,000 = 2,000

Next, examine the possibility of each allele. For the G allele, both GG and Gg individuals must be considered. Taken separately,

- GG = 2 × 600 = 1,200

- + Gg = 300

- 1,500

Bionote

The letter p is used to identify the allele frequency for the dominant allele (.75) and q for the recessive allele (.25). Note that $p + q = 1$.

The frequency for the G allele is therefore:

- 1,500/2,000 =.75

For the g allele, the calculation is similar:

- Gg = 300

- + gg = 2 × 100 = 200

- 500

The frequency for the g allele is therefore:

- 500/2,000 =.25

Hardy-Weinberg can also predict second-generation genotype frequencies. From the previous example, the allele frequencies for the only possible alleles are $p = .75(G)$ and $q = .25(g)$ after meiosis. Therefore, the probability of a GG offspring is $p \times p = p^2$ or $(.75) \times (.75) = 55$ percent. For the gg possibility, the allele frequencies are $q \times q$ or $(.25) \times (.25) = 6$ percent. For the heterozygous genotype, the dominant allele can come from either parent, so there are two possibilities: $Gg = 2pq = 2(.75)(.25) = 39$ percent.

Note that the percentages equal 100, and the allele frequencies (p and q) are identical to the genotype frequency in the first generation! Because there is no variation in this hypothetical situation, it is in Hardy-Weinberg equilibrium, and both the gene and allele frequencies will remain unchanged until acted upon by an outside force(s). Therefore, the population is in a stable equilibrium with no innate change in phenotypic characteristics. As mentioned in Historical Development and Mechanisms of Evolution and Natural Selection, populations tend not to stay in Hardy-Weinberg equilibrium for very long because of environmental pressures.

The Hardy-Weinberg equation highlights the fact that sexual reproduction does not alter the allele frequencies in a gene pool. It also helps identify a genetic equilibrium in a population that seldom exists in a natural setting because five factors impact the Hardy-Weinberg equilibrium and create their own method for *microevolution*. Note that the first four do not involve natural selection:

- Mutation
- Gene migration
- Genetic drift
- Nonrandom mating
- Natural selection

Bioterms

A mutation is an inheritable change of a gene by one of several different mechanisms that alter the DNA sequencing of an existing allele to create a new allele for that gene.

Mutation

A primary mechanism for microevolution is the formation of new alleles by *mutation*. Spontaneous errors in the replication of DNA create new alleles instantly while physical and chemical mutagens, such as ultraviolet light, create mutations constantly at a lower rate. Mutations affect the genetic equilibrium by altering the DNA, thus creating new alleles that may then become part of the reproductive gene pool for a population. When a new allele creates an advantage for the offspring, the number of individuals with the new allele may increase dramatically through successive generations. This phenomenon is not caused by the mutation somehow overmanufacturing the allele, but by the successful reproduction of individuals who possess the new allele. Because mutations are the only process that creates new alleles, it is the only mechanism that ultimately increases genetic variation.

Gene Migration

Gene migration is the movement of alleles into or out of a population either by the immigration or emigration by individuals or groups. When genes flow from one population to another, that flow may increase the genetic variation for the individual populations, but it decreases the genetic variability between the populations, making them more homogeneous. Gene migration is the opposite effect of *reproductive isolation*, which tends to be genetically near the Hardy-Weinberg equilibrium.

Genetic Drift

Genetic drift is the phenomenon whereby chance or random events change the allele frequencies in a population. Genetic drift has a tremendous effect on small populations where the gene pool is so small that minor chance events greatly influence the Hardy-Weinberg arithmetic. The failure of a single organism or small groups of organisms to reproduce creates a large genetic drift in a small population because of the loss of genes that were not conveyed to the next generation. Conversely, large populations, statistically defined as greater than 100 reproducing individuals, are proportionally less affected by isolated random events and retain more stable allele frequency with low genetic drift.

Two types of genetic drift act when a large population is modified to be considered statistically as a small population: fragmentation effect and pioneer effect.

The *fragmentation effect* is a type of genetic drift that occurs when a natural occurrence, such as a fire or hurricane, or man-made event, such as habitat destruction or overhunting, unselectively divides or reduces a population so it contains less genetic variability than the once-large population. A large population may become fragmented when a man-made dam creates a large lake where once an easily forded river provided no obstruction of movement. Likewise, a new highway can isolate species on either side. The net result is a small fragment that becomes reproductively separated from the main group. The fragmented group did not become isolated because of natural selection, so it may contain a fragment or all of the genetic variation of the larger population.

Bionote

Even when small populations recover, their genetic variability is still so low that they remain in danger of extinction from a single catastrophic event.

Zoos spend a great deal of time, money, and energy in an effort to increase their genetic diversity by locating new breeding organisms for mating, usually from other zoos.

The *pioneer effect* occurs whenever a small group breaks away from the larger population to colonize a new territory. Like the fragmentation effect, the pioneers, which may consist of only a single seed or mating pair, remain an extinction threat because they do not have the genetic diversity of the main body and therefore are less likely to produce offspring capable of surviving changes in the environment. Even though a pioneer population reproduces successfully, the gene variation has not increased. So the danger involved with survival in a changing environment still exists.

Nonrandom Mating

The Hardy-Weinberg equation assumes that all males have an equal chance to fertilize all females. However, in nature, this seldom is true because of a number of factors such as geographical proximity, as is the case in rooted plants. In fact, the ultimate non-random mating is the act of self-fertilization that is common in some plants. In other cases, as the reproductive season approaches, the number of desirable mates is limited by their presence (or absence) as well as by their competitive premating rituals. Finally, botanists and zoologists practice nonrandom mating as they attempt to breed more and better organisms for economic benefit.

Macroevolution

Macroevolution refers to evolution that occurs *above the level of species*, such as the *origin of new designs* (feathers, vertebrates from invertebrates, jaws in fish), *large scale events* (extinction of dinosaurs), *broad trends* (increase in brain size in mammals), and *major transitions* (origin of higher-level phyla). This is one of two classes of evolutionary phenomena, the other being microevolution, which refers to events and processes *at or below the level of species,* such as changes of gene frequencies in a population and speciation phenomena.

- Macroevolution: major patterns and changes among living organisms over long periods of time.
- The evidence comes from 2 main sources: fossils and comparisons between living organisms.

Macroevolution

At times, the concept of macroevolution has been defined as including evolutionary change *at and above* the level of species, and microevolution *below* the level of species. As the dividing point, the process of speciation may be viewed variously as the purview of either macroevolution or microevolution.

Macroevolution is an autonomous field of evolutionary inquiry. Paleontology, evolutionary developmental biology, comparative genomics, and molecular biology contribute many advances relating to the patterns and processes that can be classified as macroevolution.

Since the time of Darwin, the concept of macroevolution has engendered controversy. The conventional view of many evolutionists is that macroevolution is simply a continuation of microevolution on a greater scale. Others see macroevolution as more or less decoupled from microevolution. This later perspective is held both by some prominent evolutionists, as well as by many religious adherents outside the scientific community. For example, movements such as creationism and intelligent design differentiate between microevolution and macroevolution, asserting that the former (change within a species) is an observable phenomenon, but that the latter is not. Proponents of intelligent design argue that the mechanisms of evolution are incapable of giving rise to instances of specified complexity and irreducible complexity, and that while natural selection can be a creative force at the microevolutionary level, there is a divine power that is responsible as the creative force for macroevolutionary changes.

Examples of Macroevolutionary Phenomena

There are many examples of macroevolutionary phenomena found in the order Primates, including stasis, adaptive radiations, extinctions of entire lineages, co-evolution, and convergent evolution.

Adaptive radiations and stasis - Phylogenetic trees across the order Primates.

Recent studies have provided new insights about the tempo and mode of primate evolution using phylogenetic trees from genetic data gathered across the genomes of many extant primate lineages. These studies have revealed that the tempo and mode of evolution among the primates have been punctuated by the persistence of ancient relic lineages (i.e., stasis), bursts of speciation that may be consistent with adaptive radiations, and even by ongoing speciation that is governed by micro evolutionary processes. Perelman recently constructed a primate phylogenetic tree for 61 primate genera. The long branch that separates Tarsiers from other primates suggests that this group is an ancient relict lineage that has remained in stasis relative to other primates. In contrast, the Lemuriformes part of the tree has many early short branches followed by some long branches in the descendants, which suggests that the ancestors of extant lemurs experienced a rapid adaptive radiation that likely coincided with its colonization of Madagascar about 62-65 mya.

Phylogenetic trees also allow for comparing and contrasting the tempo and mode of evolution among different groups of primates inferred from fossil and genetic data. For instance, has evolution proceeded differently in New World monkeys versus Old World monkeys? New World monkeys last shared a common ancestor with Old World monkeys about 30-50 mya, but the diversification of New World monkeys and the divergence times of these lineages are not well understood. The fossil record suggests that New World monkeys have been in stasis following their initial colonization of the Americas, while Old World monkeys show evidence of 'faunal turnover' that closely matches the patterns predicted under the punctuated equilibrium model. Hodgson used molecular data to construct phylogenetic trees and to estimate divergence dates for many New World monkey species to examine the hypothesis that they have been in stasis relative to other primates. They found that New World monkeys have experienced both successive radiations and stasis during their evolution. Specifically, they found that the earliest New World monkey fossils were much older than the divergence dates they estimated for the extant New World monkey species. Using this evidence, along with patterns observed on phylogenetic trees, these researchers suggested that there was an early radiation of New World monkey ancestors followed by a period of stasis and then the extinction of most of this group prior to the Miocene. Following this period, the survivors of the original radiation then experienced a burst of rapid diversification into what would become the extant New World monkey 'crown lineages'.

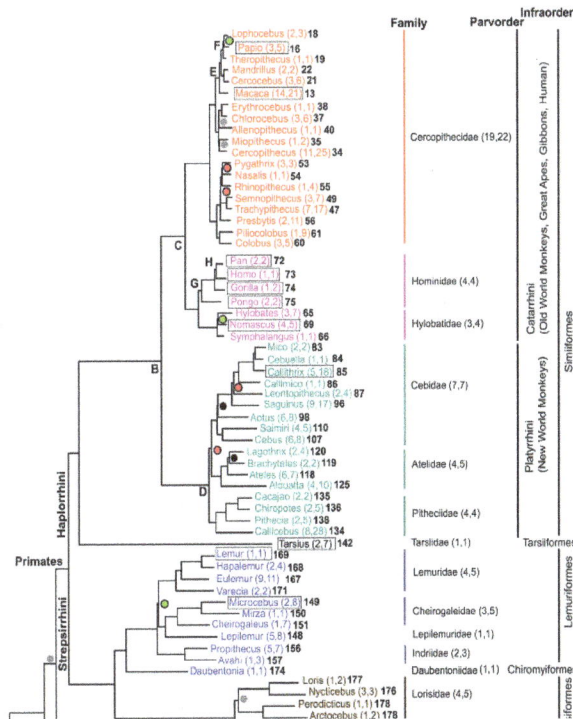

Figure: A molecular phylogeny of 61 primate genera

Adaptive radiations and extinctions - The rise and fall of Miocene Apes

Phylogenetic trees based on genetic data cannot reveal much about what might have caused adaptive radiations or extinctions. Careful examination of fossils combined with an understanding about what Earth's environment was like when these fossils were living can be used to infer what might have precipitated different macro evolutionary events. For example, during the Miocene the ancestors of Old World monkeys and apes experienced both radiations and extinctions that have been linked to climate change. In the early Miocene, primates found in Africa and the Arabian Peninsula were a diverse group that occupied tropical forests and woodlands. During the mid-Miocene, Africa reconnected with Eurasia and a major period of global warming caused the expansion of tropical habitats northward. These developments allowed the nascent hominoid lineage to branch off and colonize newly available Eurasian habitats, leading to a major proliferation of ape species across much of Eurasia. However, around 9.6 mya, a major shift to drier climates created more open habitats that led to a decline of hominoid taxa in Eurasia. By 5 mya, most ape species were extinct, except for a few that eventually led to modern-day orangutans and gibbons.

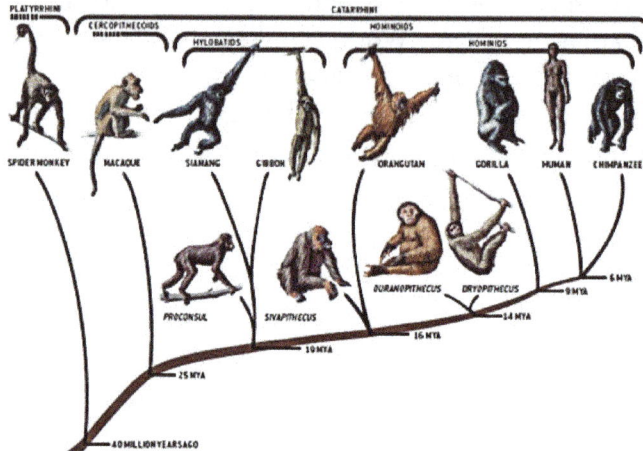

The family tree of extant hominoids includes only a small fraction of the diversity of apes that have lived on this planet. During the Miocene, up to 100 ape species once lived throughout much of Europe and Asia, but ultimately went extinct. *Proconsul* may have been the last common ancestor of extant hominoids. *Sivapithecus* was probably an ancestor to orangutans. *Ouranopithecus* or *Dryopithecus* appeared in the fossil record later in the Miocene than *Proconsul* and *Sivapithecus*. Both have been proposed as ancestors shared by all living hominoids.

Anagenesis - A gradual transition from Homo heidelbergensis to H. neanderthalensis

Understanding the tempo and mode of primate evolution is challenging because primate fossils are quite rare. Pleistocene hominins in Europe are an exception, however, since a more complete fossil record and complementary genetic data are available for this group. Comparison of Neanderthal and modern human DNA suggests that these two species shared a last common ancestor, most likely *Homo heidelbergensis*, sometime between 0.35 and 0.40 mya. These big-brained, Middle Pleistocene hominins are found in many places across the Old World and occur in the fossil record from about 1.3 mya to about 0.20 mya. They are sometimes called the 'muddle in the middle'. This group got its nickname because there are no morphological features that definitively distinguish *H. heidelbergensis* from its predecessor and its likely descendants in Africa and in Europe. This lack of diagnostic characteristics is important for understanding the tempo and mode of the later stages of human evolution. Some paleoanthropologists recognize nascent Neanderthal-like characteristics in the European branch of *H. heidelbergensis* by a gradual change over time in the fossil record towards distinctive Neanderthal traits, including large brow ridges, a slightly protruding face, and the absence of a prominent chin. These findings have led some researchers to propose that this is evidence for a continuous evolution (i.e., anagenesis) from *H. heidelbergenensis* to *H. neanderthalensis*.

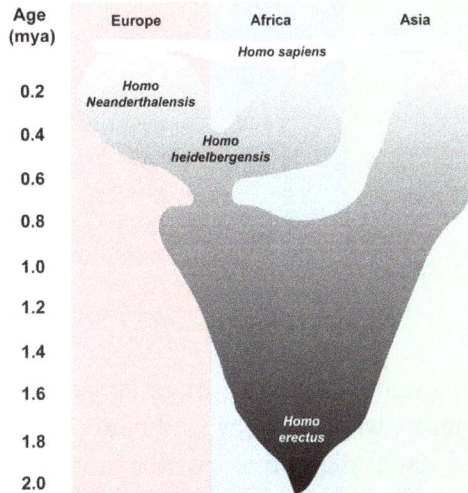

One model of Middle Pleistocene human evolution. There is evidence to suggest that a gradual transition from *Homo heidelbergensis* to *H. neanderthalensis* occurred in Europe.

Convergent evolution - Color vision in howler monkeys

A macroevolutionary perspective can also reveal patterns of convergent evolution, such as the evolution of color vision in primates. Most mammals are dichromats. This means that their vision is based on two kinds of visual pigments, or opsins. Many primates, including humans, are unique among mammals because they are trichromats who possess three types of pigments that allow them to perceive a richer array of colors compared to dichromats. There are two ways for a primate to be a trichromat. The *S* opsin is encoded by a gene located on chromosome 7, and is shared by all primates. Old World monkeys, hominoids, and humans have two additional opsin genes, located on the X chromosome, that encode pigments called '*L*' and '*M*'. New World monkeys, however, have only a single, polymorphic *M/L* opsin gene. Since the X chromosome is inherited from the mother and males are the heterogametic sex, every New World monkey male is a dichromat because he can only have a single *M* or *L* pigment along with his *S* photopigment. Females have two copies of the X chromosome, and thus, New World monkey females can be either dichromats or trichromats. Dirunal howler monkeys (*Alouatta*) are exceptions among New World primates. Both males and females can be trichromats due to a recent gene duplication of the M/L opsin gene. This 'reinvention' of trichromacy suggests that it provided them with an evolutionary advantage. This example illustrates the importance of having a macro evolutionary perspective on primate evolution because convergent evolution in color vision would not be obvious without a broad perspective on primate evolution.

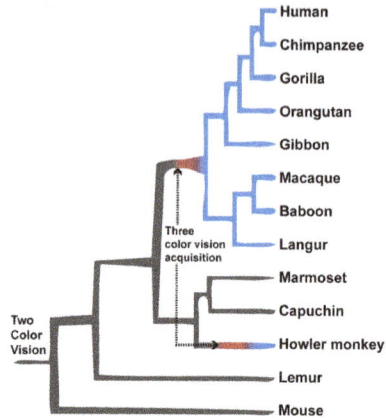

Types of color vision among major primate groups. Apes and Old World monkeys have three types of visual pigments, whereas nearly all male New World monkeys and prosimians have two types of visual pigments. Howler monkeys are unusual New World monkeys who have 're-evolved' a third visual pigment through a gene duplication event.

References

- Begon, Michael; Townsend, Colin R.; Harper, John L. (1996). Ecology: Individuals, Populations and Communities (3rd ed.). Oxford; Cambridge, MA: Blackwell Science. ISBN 978-0-632-03801-5. LCCN 95024627. OCLC 32893848

- What-are-some-examples-of-directional-selection: socratic.org, Retrieved 11 June 2018

- Negative-selection-1136: nature.com, Retrieved 31 March 2018

- Roux, Wilhelm (1881). Der Kampf der Theile im Organismus. Leipzig: Wilhelm Engelmann. OCLC 8200805. Der Kampf der Theile im Organismus on the Internet Archive Retrieved 2015-08-11

- Genetic-selection-definition-pros-cons: study.com, Retrieved 11 May 2018

- Natural-selection-genetic-drift-and-gene-flow-15186648: nature.com, Retrieved 26 March 2018

- Dennett, Daniel C. (1991). Consciousness Explained (1st ed.). Boston, MA: Little, Brown and Company. ISBN 978-0-316-18065-8. LCCN 91015614. OCLC 23648691

- Microevolution-and-macroevolution-microevolution: infoplease.com, Retrieved 24 April 2018

- Macroevolution-examples-from-the-primate-world-96679683: nature.com, Retrieved 14 April 2018

Chapter 4

Speciation

Speciation is an evolutionary process, which is responsible for the evolution of populations into distinct species. This chapter closely examines the key concepts of speciation, such as ecological speciation, allopatric speciation, peripatric speciation, parapatric speciation, cospeciation, despeciation, etc. for a holistic understanding of the subject.

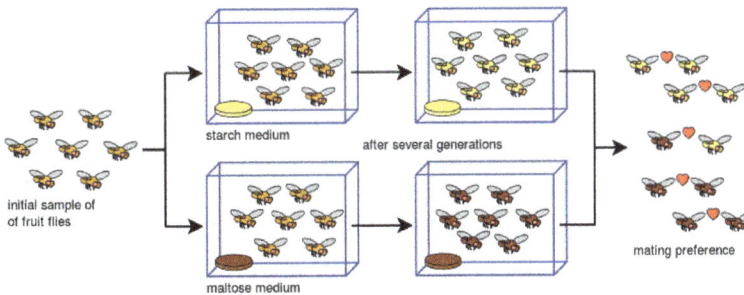

By Speciation process populations evolve to become distinct species.

The biological definition of species, which works for sexually reproducing organisms, is a group of actually or potentially interbreeding individuals. There are exceptions to this rule. Many species are similar enough that hybrid offspring are possible and may often occur in nature, but for the majority of species this rule generally holds. In fact, the presence in nature of hybrids between similar species suggests that they may have descended from a single interbreeding species, and the speciation process may not yet be completed.

Given the extraordinary diversity of life on the planet there must be mechanisms for speciation: the formation of two species from one original species. Darwin envisioned this process as a branching event and diagrammed the process in the only illustration found in *On the Origin of Species*. Compare this illustration to the diagram of elephant evolution, which shows that as one species changes over time, it branches to form more than one new species, repeatedly, as long as the population survives or until the organism becomes extinct.

For speciation to occur, two new populations must be formed from one original population and they must evolve in such a way that it becomes impossible for individuals from the two new populations to interbreed. Biologists have proposed mechanisms by which this could occur that fall into two broad categories. Allopatric speciation (*allo–* = "other"; *–patric* = "homeland") involves geographic separation of populations from a parent species and subsequent evolution. Sympatric speciation (*sym–* = "same"; *–patric* = "homeland") involves speciation occurring within a parent species remaining in one location.

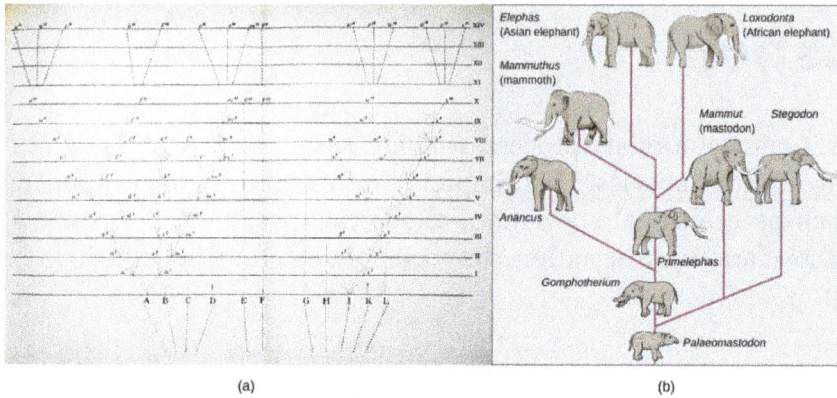

(a)   (b)

The only illustration in Darwin's *On the Origin of Species* is (a) a diagram showing speciation events leading to biological diversity. The diagram shows similarities to phylogenetic charts that are drawn today to illustrate the relationships of species. (b) Modern elephants evolved from the *Palaeomastodon*, a species that lived in Egypt 35–50 million years ago.

Biologists think of speciation events as the splitting of one ancestral species into two descendant species. There is no reason why there might not be more than two species formed at one time except that it is less likely and multiple events can be conceptualized as single splits occurring close in time.

Adaptive Radiation

In some cases, a population of one species disperses throughout an area, and each finds a distinct niche or isolated habitat. Over time, the varied demands of their new lifestyles lead to multiple speciation events originating from a single species. This is called adaptive radiation because many adaptations evolve from a single point of origin; thus, causing the species to radiate into several new ones. Island archipelagos like the Hawaiian Islands provide an ideal context for adaptive radiation events because water surrounds each island which leads to geographical isolation for many organisms. The Hawaiian honeycreeper illustrates one example of adaptive radiation.

Reproductive Isolation

Given enough time, the genetic and phenotypic divergence between populations will affect characters that influence reproduction: if individuals of the two populations were to be brought together, mating would be less likely, but if mating occurred, offspring would be non-viable or infertile. Many types of diverging characters may affect the reproductive isolation, the ability to interbreed, of the two populations.

Reproductive isolation can take place in a variety of ways. Scientists organize them into two groups: Prezygotic barriers and postzygotic barriers. Recall that a zygote is a fertilized egg: The first cell of the development of an organism that reproduces sexually. Therefore, a prezygotic barrier is a mechanism that blocks reproduction from taking place; this includes barriers that prevent fertilization when organisms attempt reproduction. A postzygotic barrier occurs after zygote formation; this includes organisms that don't survive the embryonic stage and those that are born sterile.

Some types of prezygotic barriers prevent reproduction entirely. Many organisms only reproduce at certain times of the year, often just annually. Differences in breeding schedules, called temporal isolation, can act as a form of reproductive isolation. For example, two species of frogs inhabit the same area, but one reproduces from January to March, whereas the other reproduces from March to May.

Modes

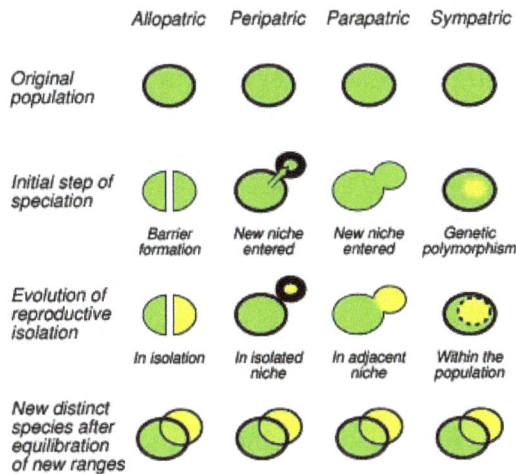

Comparison of allopatric, peripatric, parapatric and sympatric speciation

All forms of natural speciation have taken place over the course of evolution; however, debate persists as to the relative importance of each mechanism in driving biodiversity.

One example of natural speciation is the diversity of the three-spined stickleback, a marinefish that, after the last glacial period, has undergone speciation into new freshwater colonies in isolated lakes and streams. Over an estimated 10,000 generations, the sticklebacks show structural differences that are greater than those seen between different genera of fish including variations in fins, changes in the number or size of their bony plates, variable jaw structure, and color differences.

Allopatric

During allopatric speciation, a population splits into two geographically isolated populations (for example, by habitat fragmentation due to geographical change such as mountain formation). The isolated populations then undergo genotypic or phenotypic divergence as:

(a) They become subjected to dissimilar selective pressures;

(b) They independently undergo genetic drift;

(c) Different mutations arise in the two populations.

When the populations come back into contact, they have evolved such that they are reproductively isolated and are no longer capable of exchanging genes. Island genetics is the term associated with the tendency of small, isolated genetic pools to produce unusual traits. Examples include insular dwarfism and the radical changes among certain famous island chains, for example on Komodo. The Galápagos Islands are particularly famous for their influence on Charles Darwin. During his five weeks there he heard that Galápagos tortoises could be identified by island, and noticed that finches differed from one island to another, but it was only nine months later that he reflected that such facts could show that species were changeable. When he returned to England, his speculation on evolution deepened after experts informed him that these were separate species, not just varieties, and famously that other differing Galápagos birds were all species of finches. Though the finches were less important for Darwin, more recent research has shown the birds now known as Darwin's finches to be a classic case of adaptive evolutionary radiation.

Peripatric

In peripatric speciation, a sub form of allopatric speciation, new species are formed in isolated, smaller peripheral populations that are prevented from exchanging genes with the main population. It is related to the concept of a founder effect, since small populations often undergo bottlenecks. Genetic drift is often proposed to play a significant role in peripatric speciation.

Case studies include Mayr's investigation of bird fauna; the Australian bird *Petroica multicolor*; and reproductive isolation in populations of *Drosophila* subject to population bottlenecking.

Parapatric

In parapatric speciation, there is only partial separation of the zones of two diverging populations afforded by geography; individuals of each species may come in contact or cross habitats from time to time, but reduced fitness of the heterozygote leads to selection for behaviours or mechanisms that prevent their interbreeding. Parapatric speciation is modelled on continuous variation within a "single," connected habitat acting as a source of natural selection rather than the effects of isolation of habitats produced in peripatric and allopatric speciation.

Parapatric speciation may be associated with differential landscape-dependent selection. Even if there is a gene flow between two populations, strong differential selection may impede assimilation and different species may eventually develop. Habitat differences may be more important in the development of reproductive isolation than the isolation time. Caucasian rock lizards *Darevskia rudis*, *D. valentini* and *D. portschinskii* all hybridize with each other in their hybrid zone; however, hybridization is stronger between *D. portschinskii* and *D. rudis*, which separated earlier but live in similar habitats than between *D. valentini* and two other species, which separated later but live in climatically different habitats.

Ecologists refer to parapatric and peripatric speciation in terms of ecological niches. A niche must be available in order for a new species to be successful. Ring species such as *Larus* gulls have been claimed to illustrate speciation in progress, though the situation may be more complex. The grass *Anthoxanthum odoratum* may be starting parapatric speciation in areas of mine contamination.

Sympatric

Cichlids such as *Haplochromis nyererei* diversified by sympatric speciation in the Rift Valley lakes.

Sympatric speciation is the formation of two or more descendant species from a single ancestral species all occupying the same geographic location.

Often-cited examples of sympatric speciation are found in insects that become dependent on different host plants in the same area.

The best illustrated example of sympatric speciation is that of the cichlids of East Africa inhabiting the Rift Valley lakes, particularly Lake Victoria, Lake Malawi and Lake Tanganyika. There are over 800 described species, and according to estimates, there could be well over 1,600 species in the region. Their evolution is cited as an example of both natural and sexual selection. A 2008 study suggests that sympatric speciation has occurred in Tennessee cave salamanders. Sympatric speciation driven by ecological factors may also account for the extraordinary diversity of crustaceans living in the depths of Siberia's Lake Baikal.

Budding speciation has been proposed as a particular form of sympatric speciation, whereby small groups of individuals become progressively more isolated from the ancestral stock by breeding preferentially with one another. This type of speciation would be driven by the conjunction of various advantages of inbreeding such as the expression of advantageous recessive phenotypes, reducing the recombination load, and reducing the cost of sex.

Rhagoletis pomonella, the hawthorn fly, appears to be in the process of sympatric speciation.

The hawthorn fly (*Rhagoletis pomonella*), also known as the apple maggot fly, appears to be undergoing sympatric speciation. Different populations of hawthorn fly feed on different fruits. A distinct population emerged in North America in the 19th century some time after apples, a non-native species, were introduced. This apple-feeding population normally feeds only on apples and not on the historically preferred fruit of hawthorns. The current hawthorn feeding population does not normally feed on apples. Some evidence, such as that six out of thirteen allozyme loci are different, that hawthorn flies mature later in the season and take longer to mature than apple flies; and that there is little evidence of interbreeding (researchers have documented a 4-6% hybridization rate) suggests that sympatric speciation is occurring.

Methods of Selection

Reinforcement

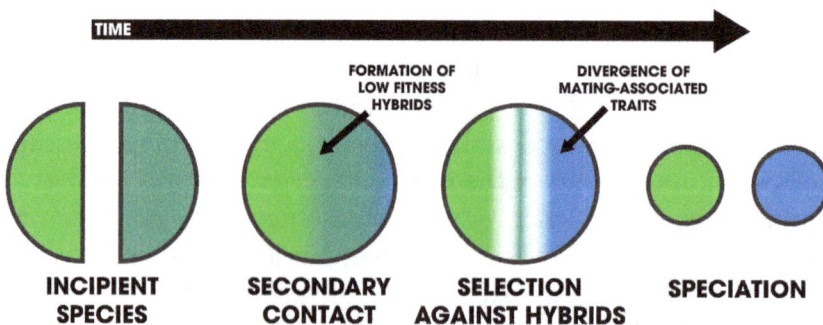

Reinforcement assists speciation by selecting against hybrids.

Reinforcement, sometimes referred to as the Wallace effect, is the process by which natural selection increases reproductive isolation. It may occur after two populations of the same species are separated and then come back into contact. If their reproductive isolation was complete, then they will have already developed into two separate incompatible species. If their reproductive isolation is incomplete, then further mating between the populations will produce hybrids, which may or may not be fertile. If the hybrids are infertile, or fertile but less fit than their ancestors, then there will be further reproductive isolation and speciation has essentially occurred (e.g., as in horses and donkeys).

The reasoning behind this is that if the parents of the hybrid offspring each have naturally selected traits for their own certain environments, the hybrid offspring will bear traits from both, therefore would not fit either ecological niche as well as either parent. The low fitness of the hybrids would cause selection to favor assortative mating, which would control hybridization. This is sometimes called the Wallace effect after the evolutionary biologist Alfred Russel Wallace who suggested in the late 19th century that it might be an important factor in speciation.

Conversely, if the hybrid offspring are more fit than their ancestors, then the populations will merge back into the same species within the area they are in contact.

Reinforcement favoring reproductive isolation is required for both parapatric and sympatric speciation. Without reinforcement, the geographic area of contact between different forms of the same species, called their "hybrid zone," will not develop into a boundary between the different species. Hybrid zones are regions where diverged populations meet and interbreed. Hybrid offspring are very common in these regions, which are usually created by diverged species coming into secondary contact. Without reinforcement, the two species would have uncontrollable inbreeding.

Ecological

Ecological selection is "the interaction of individuals with their environment during resource acquisition". Natural selection is inherently involved in the process of speciation, whereby, "under ecological speciation, populations in different environments, or populations exploiting different resources, experience contrasting natural selection pressures on the traits that directly or indirectly bring about the evolution of reproductive isolation". Evidence for the role ecology plays in the process of speciation exists. Studies of stickleback populations support ecologically-linked speciation arising as a by-product, alongside numerous studies of parallel speciation, where isolation evolves between independent populations of species adapting to contrasting environments than between independent populations adapting to similar environments. Ecological speciation occurs with much of the evidence, "accumulated from top-down studies of adaptation and reproductive isolation".

Sexual Selection

It is widely appreciated that sexual selection could drive speciation in many clades, independently of natural selection. However the term "speciation", in this context, tends to be used in two different, but not mutually exclusive senses. The first and most commonly used sense refers to the "birth" of new species. That is, the splitting of an existing species into two separate species, or the budding off of a new species from a parent species, both driven by a biological "fashion fad" (a preference for a feature, or features, in one or both sexes, that do not necessarily have any adaptive qualities). In the second sense, "speciation" refers to the wide-spread tendency of sexual creatures to be grouped into clearly defined species, rather than forming a continuum of phenotypes both in time and space - which would be the more obvious or logical consequence of natural selection. This was indeed recognized by Darwin as problematic, and included in his *On the Origin of Species*, under the heading "Difficulties with the Theory". There are several suggestions as to how mate choice might play a significant role in resolving Darwin's dilemma.

Artificial Speciation

Gaur (Indian bison) can interbreed with domestic cattle.

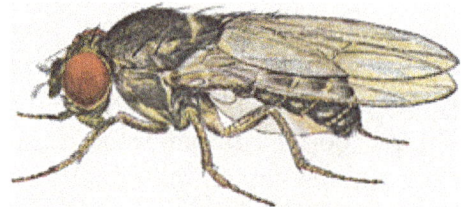

Male *Drosophila pseudoobscura*

New species have been created by animal husbandry, but the dates and methods of the initiation of such species are not clear. Often, the domestic counterpart of the wild ancestor can still interbreed and produce fertile offspring as in the case of domestic cattle, that can be considered the same species as several varieties of wild ox, gaur, yak, etc., or domestic sheep that can interbreed with the mouflon.

The best-documented creations of new species in the laboratory were performed in the late 1980s. William R. Rice and George W. Salt bred *Drosophila melanogaster* fruit flies using a maze with three different choices of habitat such as light/dark and wet/dry. Each generation was placed into the maze, and the groups of flies that came out of two of the eight exits were set apart to breed with each other in their respective groups. After thirty-five generations, the two groups and their offspring were isolated reproductively because of their strong habitat preferences: they mated only within the areas they preferred, and so did not mate with flies that preferred the other areas. The history of such attempts is described by Rice and Elen E. Hostert. Diane Dodd used a laboratory experiment to show how reproductive isolation can evolve in *Drosophila pseudoobscura* fruit flies after several generations by placing them in different media, starch- and maltose-based media.

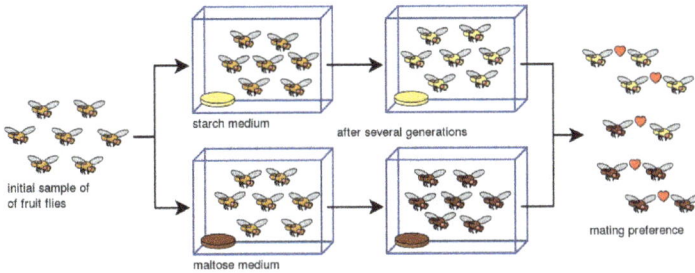

Dodd's experiment has been easy for many others to replicate, including with other kinds of fruit flies and foods. Research in 2005 has shown that this rapid evolution of reproductive isolation may in fact be a relic of infection by *Wolbachia* bacteria.

Alternatively, these observations are consistent with the notion that sexual creatures are inherently reluctant to mate with individuals whose appearance or behavior is different from the norm. The risk that such deviations are due to heritable maladaptations is very high. Thus, if a sexual creature, unable to predict natural selection's future direction, is conditioned to produce the fittest offspring possible, it will avoid mates with unusual habits or features. Sexual creatures will then inevitably tend to group themselves into reproductively isolated species.

Genetics

Few speciation genes have been found. They usually involve the reinforcement process of late stages of speciation. In 2008, a speciation gene causing reproductive isolation was reported. It causes hybrid sterility between related subspecies. The order of speciation of three groups from a common ancestor may be unclear or unknown; a collection of three such species is referred to as a "trichotomy."

Speciation via Polyploidy

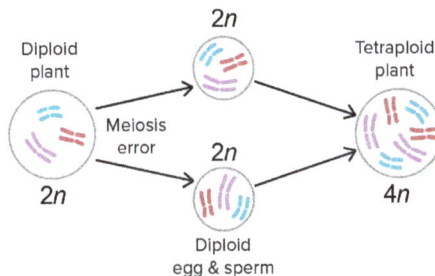

Speciation via polyploidy: A diploidcell undergoes failed meiosis, producing diploid gametes, which self-fertilize to produce a tetraploid zygote. In plants, this can effectively be a new species, reproductively isolated from its parents, and able to reproduce.

Polyploidy is a mechanism that has caused many rapid speciation events in sympatry because offspring of, for example, tetraploid x diploid matings often result in triploid sterile progeny. However, not all polyploids are reproductively isolated from their parental plants, and gene flow may still occur for example through triploid hybrid x diploid matings that produce tetraploids, or matings between meiotically unreduced gametes from diploids and gametes from tetraploids. It has been suggested that many of the existing plant and most animal species have undergone an event of polyploidization in their evolutionary history. Reproduction of successful polyploid species is sometimes asexual, by parthenogenesis or apomixis, as for unknown reasons many asexual organisms are polyploid. Rare instances of polyploid mammals are known, but most often result in prenatal death.

Hybrid Speciation

Hybridization between two different species sometimes leads to a distinct phenotype. This phenotype can also be fitter than the parental lineage and as such natural selection may then favor these individuals. Eventually, if reproductive isolation is achieved, it may lead to a separate species. However, reproductive isolation between hybrids and their parents is particularly difficult to achieve and thus hybrid speciation is considered an extremely rare event. The Mariana mallard is thought to have arisen from hybrid speciation.

Hybridization is an important means of speciation in plants, since polyploidy (having more than two copies of each chromosome) is tolerated in plants more readily than in animals. Polyploidy is important in hybrids as it allows reproduction, with the two different sets of chromosomes each being able to pair with an identical partner during meiosis. Polyploids also have more genetic diversity, which allows them to avoid inbreeding depression in small populations.

Hybridization without change in chromosome number is called homoploid hybrid speciation. It is considered very rare but has been shown in *Heliconius* butterflies and sunflowers. Polyploid speciation, which involves changes in chromosome number, is a more common phenomenon, especially in plant species.

Gene Transposition

Theodosius Dobzhansky, who studied fruit flies in the early days of genetic research in 1930s, speculated that parts of chromosomes that switch from one location to another might cause a species to split into two different species. He mapped out how it might be possible for sections of chromosomes to relocate themselves in a genome. Those mobile sections can cause sterility in inter-species hybrids, which can act as a speciation pressure. In theory, his idea was sound, but scientists long debated whether it actually happened in nature. Eventually a competing theory involving the gradual accumulation of mutations was shown to occur in nature so often that geneticists largely dismissed the moving gene hypothesis. However, 2006 research shows that jumping of a gene from one chromosome to another can contribute to the birth of new species. This validates the reproductive isolation mechanism, a key component of speciation.

Rates

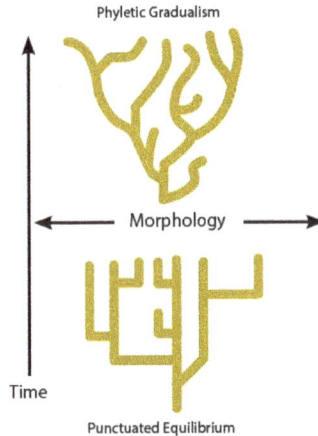

Phyletic gradualism, above, consists of relatively slow change over geological time.
Punctuated equilibrium, bottom, consists of morphological stability and rare,
relatively rapid bursts of evolutionary change.

There is debate as to the rate at which speciation events occur over geologic time. While some evolutionary biologists claim that speciation events have remained relatively constant and gradual over time, some palaeontologists such as Niles Eldredge and Stephen Jay Gould have argued that species usually remain unchanged over long stretches of time, and that speciation occurs only over relatively brief intervals, a view known as *punctuated equilibrium*.

Punctuated Evolution

Evolution can be extremely rapid, as shown in the creation of domesticated animals and plants in a very short geological space of time, spanning only a few tens of thousands of years. Maize (*Zea mays*), for instance, was created in Mexico in only a few thousand years, starting about 7,000 to 12,000 years ago. This raises the question of why the long term rate of evolution is far slower than is theoretically possible.

Plants and domestic animals can differ markedly from their wild ancestors:

Top: wild teosinte; middle: maize-teosinte hybrid; bottom: maize

Ancestral wild cabbage

Domesticated cauliflower

Ancestral Prussian carp

Domestic goldfish

Ancestral mouflon

Domestic sheep

Evolution is imposed on species or groups. It is not planned or striven for in some Lamarckist way. The mutations on which the process depends are random events, and, except for the "silent mutations" which do not affect the functionality or appearance of the carrier, are thus usually disadvantageous, and their chance of proving to be useful in the future is vanishingly small. Therefore, while a species or group might benefit from being able to adapt to a new environment by accumulating a wide range of genetic variation, this is to the detriment of the *individuals* who have to carry these mutations until a small, unpredictable minority of them ultimately contributes to such an adaptation. Thus, the *capability* to evolve would require group selection, a concept discredited by (for example) George C. Williams, John Maynard Smith and Richard Dawkins as selectively disadvantageous to the individual.

The resolution to Darwin's second dilemma might thus come about as follows:

If sexual individuals are disadvantaged by passing mutations on to their offspring, they will avoid mutant mates with strange or unusual characteristics. Mutations that affect the external appearance of their carriers will then rarely be passed on to the next and subsequent generations. They would therefore seldom be tested by natural selection. Evolution is, therefore, effectively halted or slowed down considerably. The only mutations that can accumulate in a population, on this punctuated equilibrium view, are ones that have no noticeable effect on the outward appearance and functionality of their bearers (i.e., they are "silent" or "neutral mutations," which can be, and are, used to trace the relatedness and age of populations and species.) This argument implies that evolution can only occur if mutant mates cannot be avoided, as a result of a severe scarcity of potential mates. This is most likely to occur in small, isolated communities. These occur most commonly on small islands, in remote valleys, lakes, river systems, or caves, or during the aftermath of a mass extinction. Under these circumstances, not only is the choice of mates severely restricted but population bottlenecks, founder effects, genetic drift and inbreeding cause rapid, random changes in the isolated population's genetic composition. Furthermore, hybridization with a related species trapped in the same isolate might introduce additional genetic changes. If an isolated population such as this survives its genetic upheavals, and subsequently expands into an unoccupied niche, or into a niche in which it has an advantage over its competitors, a new species, or subspecies, will have come in being. In geological terms this will be an abrupt event. A resumption of avoiding mutant mates will thereafter result, once again, in evolutionary stagnation.

In apparent confirmation of this punctuated equilibrium view of evolution, the fossil record of an evolutionary progression typically consists of species that suddenly appear, and ultimately disappear, hundreds of thousands or millions of years later, without any change in external appearance. Graphically, these fossil species are represented by lines parallel with the time axis, whose lengths depict how long each of them existed. The fact that the lines remain parallel with the time axis illustrates the unchanging appearance of each of the fossil species depicted on the graph. During each species' existence new species appear at random intervals, each also lasting many hundreds of thousands of years before disappearing without a change in appearance. The exact relatedness of these concurrent species is generally impossible to determine. This is illustrated in the diagram depicting the distribution of hominin species through time since the hominins separated from the line that led to the evolution of our closest living primate relatives, the chimpanzees.

For similar evolutionary time lines see, for instance, the paleontological list of African dinosaurs, Asian dinosaurs, the Lampriformes and Amiiformes.

Ecological Speciation

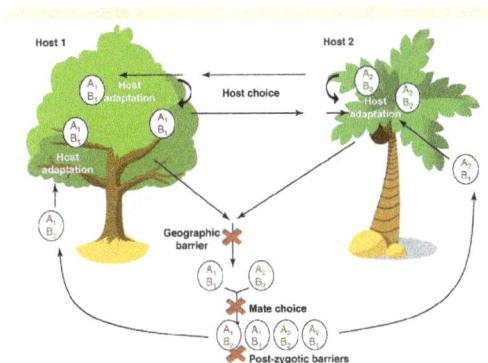

'Ecological speciation' is a mode of speciation in which adaptation to different environments causes populations to diverge and develop reproductive isolation from each other. This classical hypothesis has matured in the field of evolutionary ecology with the accumulation of new empirical data. A full understanding of the mechanisms by which ecological speciation leads to biological diversification requires the synthesis of ideas from many different disciplines, including evolution, ecology, and genetics.

Ecological speciation occurs when adaptation to divergent environments, such as different resources or habitats, leads to the evolution of reproductive isolation. More specifically, divergent (or disruptive) selection between environments causes the adaptive divergence of populations, which leads to the evolution of reproductive barriers that decrease, and ultimately cease, gene flow. Supported by a growing number of specific examples, ecological speciation is thought to be a primary driving force in evolutionary diversification, exemplified most obviously in adaptive radiations.

As acceptance of the importance of ecological speciation has grown, so too has the recognition that it is not all powerful. Specifically, a number of instances of nonecological speciation and nonadaptive radiation seem likely, and colonization of different environments does not always lead to speciation. This latter point is obvious when one recognizes that although essentially all species are composed of a number of populations occupying divergent environments, only a fraction of these ever spin off to become full-fledged species. Instead, populations occupying divergent environments or using different resources show varying levels of progress toward ecological speciation—and this variation provides the substrate to study factors that promote and constrain progress along the speciation continuum.

Models for Progress Toward Ecological Speciation

D. J. Funk addresses this topic by first clarifying the relationship between sympatric speciation (whereby reproductively isolated populations evolve from an initially panmictic population) and ecological speciation (whereby reproductive isolation evolves as a consequence of divergent/disruptive natural selection). These are orthogonal concepts. First, even if disruptive selection is a common way of achieving sympatric speciation, this can also be caused by other factors, such as changes in chromosome number. Second, ecological speciation can readily occur in allopatry. Funk then introduces four new concepts aiming to reduce confusion in the literature. Sympatric race is a generalization of host race (usually used for herbivores or parasites) and refers to any sympatric populations that experience divergent selection and are partly but incompletely reproductively isolated. Envirotypes are populations that differ due to phenotypic plasticity. Host forms are populations that exhibit host-associated variation, but for which the nature of variation (e.g., envirotype, host race, cryptic species) has not yet been diagnosed. Ecological forms are a generalization of host forms for nonherbivore or parasitic taxa. The two latter concepts acknowledge the fact that one has an incomplete understanding of speciation. To overcome the problem of overdiagnosing host races, Funk introduces five criteria, based on host association and choice, coexistence pattern, genetic differentiation, mate choice, gene flow, and hybrid unfitness. Funk's maple and willow associated phytophagous populations of Neochlamisus bebbianae leaf beetle meet all these criteria and can, therefore, be considered as host races.

Another phytophagy-inspired conceptual model for how an insect species initially using one plant species might diversify into multiple insect species using different host plants is presented by S. Heard. This effort explicitly links variation in host plant use within insect species or races to the formation of different host races and species. In this proposed "gape-and-pinch" model, Heard posits four stages (or "hypotheses") of diversification defined in part by overlap in the plant trait space used by the insect races/species. In the first stage "adjacent errors," some individuals within an insect species using one plant species might "mistakenly" use individuals of another plant species that have similar trait values to their normal host plant species. In the next stage "adjacent oligophagy," populations formed by the insects that shifted plant species then experience divergent selection—and undergo adaptive divergence—leading to a better use of that new host. In the third stage "trait distance-divergence," competition and reproductive interactions cause character displacement between the emerging insect races or species so that they become specialized on particularly divergent subsets of the trait distributions of the two plant species. In the final stage "distance relaxation," the new species become so divergent that they no longer interact, and can then evolve to use trait values more typical of each plant species. Heard provides a theoretical and statistical framework for testing this model and applies it to insects using goldenrod plants.

Local adaptation is often the first step in ecological speciation, and so factors influencing local adaptation will be critical for ecological speciation. Local adaptation can either increase over time (if more specialized alleles spread), eventually leading to speciation, or it can decrease over time (if more generalist alleles spread). T. Lenormand reviews the conditions that favor these different scenarios and emphasizes the role of three positive feedback loops that favor increased specialization. In the demographic loop, local adaptation results in higher population density, which in turn favors the recruitment of new locally adapted alleles. In the recombination loop, locally adapted alleles are more likely to be recruited in genomic regions already harbouring loci with locally adapted alleles, thereby generating genomic regions of particular importance to local adaptation. In the reinforcement loop, local adaptation selects for traits that promote premating isolation (reinforcement), which in turn increases the recruitment and frequency of locally adapted alleles. Lenormand then details the mechanisms involved in reinforcement, particularly assortative mating, dispersal, and recombination. He highlights that these characteristics represent the three fundamental steps in a sexual life cycle (syngamy, dispersal, and meiosis) and that they promote genetic clustering at several levels (within locus, among individuals, among loci). His new classification is orthogonal to, and complements, the traditional one-versus two-allele distinction. Overall, the rates of increased specialization and reinforcement determine progress toward ecological speciation.

One of the major constraints on ecological speciation is the establishment of self-sustaining populations in new/marginal environments, because the colonizing individuals are presumably poorly adapted to the new conditions. This difficulty might be eased through facilitation, the amelioration of habitat conditions by the presence of neighbouring living organisms (biotic components). According to this process, the benefactor's "environmental bubble" facilitates the beneficiary's adaptation to marginal conditions, which can result in ecological speciation if gene flow from the core habitat is further reduced. At the same time, however, facilitation might hinder further progress toward ecological speciation by maintaining gene flow between environments and by preventing reinforcement in secondary contact zones.

Variable Progress toward Ecological Speciation in Nature

Three spine stickleback fish, with their diverse populations adapted to different habitats, had provided a number of examples of how adaptive divergence can promote ecological speciation.

Another classic system for studying ecological speciation, or more generally adaptive radiation, is Anolis lizards of the Caribbean. In particular, many of the larger islands contain repeated radiations of similar "ecomorph" species in similar habitats. Contrasting with this predictable and repeatable diversity on large islands, smaller islands contain only a few species. Y. Surget-Groba, H. Johansson, and R. S. Thorpe studied populations of Anolis roquet from Martinique. This species contains populations with divergent mitochondrial lineages, a consequence of previous allopatric episodes, and is distributed over a range of habitats. It can, therefore, be used to address the relative importance of past allopatry, present ecological differences, and their combination in determining progress toward ecological speciation.

Factors Affecting Progress toward Ecological Speciation

The Role of Pollinators/Parasites

In many cases of ecological speciation, we think of the populations in question colonizing and adapting to divergent environments/resources, such as different plants or other food types. However, environments can also "colonize" the populations in question that might then speciate as a result. Colonization by different pollinators and subsequent adaptation to them, for example, is expected to be particularly important for angiosperms. A particularly spectacular example involves sexually deceptive Orchids, where flowers mimic the scent and the appearance of female insects and are then pollinated during attempted copulation by males.

Parasites can be thought of as another instance of different environments "colonizing" a focal species and then causing divergent/disruptive selection and (perhaps) ecological speciation.

Differences in parasites could contribute to ecological speciation in three major ways. First, divergent parasite communities could cause selection against locally adapted hosts that move between those communities, as well as any hybrids. Second, adaptation to divergent parasite communities could cause assortative mating to evolve as a pleiotropic by-product, such as through divergence in MHC genotypes that are under selection by parasites and also influence mate choice. Third, sexual selection might lead females in a given population to prefer males that are better adapted to local parasites and can thus achieve better condition.

The Role of Habitat Choice

The importance of habitat (or host) isolation in ecological speciation is widely recognized. This habitat isolation is determined by habitat choice (preference or avoidance), competition, and habitat performance (fitness differences between habitats.

Role of habitat choice can be based on three largely independent criteria:

(1) Whether habitat choice allows the establishment of a stable polymorphism maintained by selection without interfering with mating randomness or if it also promotes assortative mating;

(2) Whether it involves one-allele or two-allele mechanisms of inheritance;

(3) Whether traits are of single or multiple effect, the latter when habitat choice is simultaneously under direct selection and contributes to assortative mating.

The Role of Phenotypic Plasticity

Phenotypic plasticity, the ability of a single genotype to express different phenotypes under different environmental conditions, has long been seen as an alternative to genetic divergence, and therefore as potential constraint on adaptive evolution. More recently, however, adaptive phenotypic plasticity has been rehabilitated as a factor potentially favoring divergent evolution by enabling colonizing new niches, where divergent selection can then act on standing genetic variation.

Allopatric Speciation

Allopatric speciation is speciation that happens when two populations of the same species become isolated from each other due to geographic changes. Speciation is a gradual process by which populations evolve into different species. A species is itself defined as a population that can interbreed, so during speciation, members of a population form two or more distinct populations that can no longer breed with each other.

Steps of Allopatric Speciation

1. A geographic change separates members of a population into more than one group. Such changes could include the formation of a new mountain range or new waterway, or the development of new canyons, for example. Also, human activities such as civil engineering, agriculture, and pollution can have an effect on habitable environments and cause some members of a population to migrate.

2. Different gene mutations occur and build up in the different populations over time. The different variations of genes may lead to different characteristics between the two populations.

3. The populations become so different that members of the different populations can no longer breed with each other anymore if were they to be in the same habitat in the same time. If this is the case, allopatric speciation has occurred.

The following diagram represents an experiment on fruit flies where the population was forcibly separated and the two groups were fed a different diet. After many generations the flies looked different and preferred to mate with flies from their own group. If these two populations continued to diverge for a long time, they could become two different species through allopatric speciation.

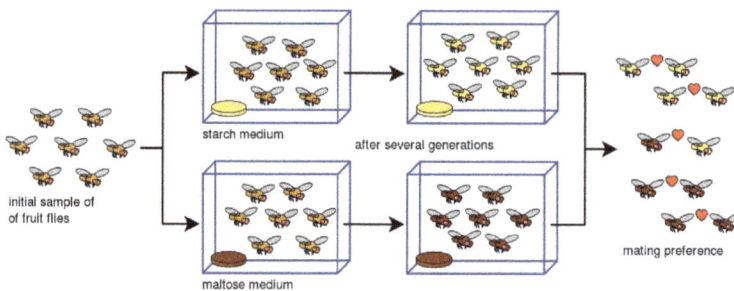

starch medium   after several generations

initial sample of
of fruit flies

mating preference

maltose medium

Examples of Allopatric Speciation

Darwin's Finches

A major example of allopatric speciation occurred in the Galapagos finches that Charles Darwin studied. There are about 15 different species of finches on the Galapagos islands, and they each look different and have specialized beaks for eating different types of foods, such as insects, seeds, and flowers. All of these finches came from a common ancestor species that must have emigrated to the different islands. Once populations were established on the islands, they became isolated from each other and different mutations arose. The mutations that caused the birds to be most successful in their respective environments became more and more prevalent, and many different species formed over time. When many new species emerge from one common ancestor in a relatively quickly geological timeframe, this is called adaptive radiation.

When the Grand Canyon was formed, it created a natural barrier between the squirrels living in the area. About 10,000 years ago, the squirrel population was separated from each other by this geographic change and could no longer live in the same area. Over thousands of years, the divided squirrel populations became two different species. Kaibab squirrels live on the north rim of the canyon and have a small range, while Abert squirrels live on the south rim and live in a much larger range. Members of these two species have a similar size, shape, and diet, and slight color differences, but they are no longer in contact with each other and have become so different during their separation that they are now separate species.

Parapatric Speciation

In parapatric speciation a species is spread out over a large geographic area. Although it is possible for any member of the species to mate with another member, individuals only mate with those in their own geographic region. Like allopatric and peripatric speciation, different habitats influence the development of different species in parapatric speciation. Instead of being separated by a physical barrier, the species are separated by differences in the same environment.

Parapatric speciation sometimes happens when part of an environment has been polluted. Mining activities leave waste with high amounts of metals like lead and zinc. These metals are absorbed into the soil, preventing most plants from growing. Some grasses, such as buffalo grass, can tolerate the metals. Buffalo grass, also known as vanilla grass, is native to Europe and Asia, but is now found throughout North and South America, too. Buffalo grass has become a unique species from the grasses that grow in areas not polluted by metals. Long distances can make it impractical to travel to reproduce with other members of the species. Buffalo grass seeds pass on the characteristics of the members in that region to offspring. Sometimes a species that is formed by parapatric speciation is especially suited to survive in a different kind of environment than the original species.

Peripatric Speciation

Peripatric speciation is a form of speciation, the formation of new species through evolution. In this form, new species are formed in isolated peripheral populations; this is similar to allopatric speciation in that populations are isolated and prevented from exchanging genes. However, peripatric speciation, unlike allopatric speciation, proposes that one of the populations is much smaller than the other. One possible consequence of peripatric speciation is that a geographically widespread ancestral species becomes paraphyletic, thereby becoming a paraspecies. The concept of a para species is therefore a logical consequence of the Evolutionary Species Concept, by which one species give rise to a daughter species. The evolution of the polar bear from the brown bear is a well-documented example of a living species that gave rise to another living species through the evolution of a population located at the margin of the ancestral species' range.

Peripatric speciation

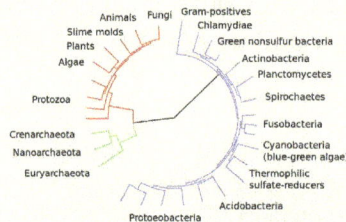

Peripatric speciation was originally proposed by Ernst Mayr, and is related to the founder effect, because small living populations may undergo selection bottlenecks. The founder-effect is based on models that suggest peripatric speciation can occur by the interaction of selection and genetic drift; of which is often proposed to play a significant role in peripatric speciation.

Theoretical Framework

Theoretical models of peripatric speciation are identical to models of vicariance(allopatry); however, the nature of dispersal and colonization into novel environments may give rise to rapid speciation events. In addition, individuals colonizing a new habitat likely contain only a small "sample" of the genetic variation of the original population; thus promoting divergence due to selective pressures, and possibly leading to the rapid fixation of an allele within the descendant population. This gives rise to the potential for genetic incompatibilities to evolve.

Laboratory Evidence

Peripatric speciation has been researched in both laboratory studies and nature. Coyne and Orr in *Speciation* suggest that most laboratory studies of allopatric speciation are also examples of peripatric speciation due to their small population sizes and the inevitable divergent selection that they undergo.

Much of the laboratory research concerning peripatry is inextricably linked to founder-effect research. Coyne and Orr conclude that selection's role in speciation is well established, whereas genetic drift's role is unsupported by experimental and field data—suggesting that founder-effect speciation does not occur. Nevertheless, a great deal of research has been conducted on the matter, and one study conducted involving bottleneck populations of *Drosophila pseudoobscura* found evidence of isolation after a single bottleneck.

Observational Evidence

Island species provide direct evidence of speciation occurring peripatrically in such that, "the presence of endemic species on oceanic islands whose closest relatives inhabit a nearby continent" must have originated by a colonization event.

Drosophila species on the Hawaiian archipelago have helped researchers understand speciation processes in great detail. It is well established that *Drosophila* has undergone an adaptive radiation into hundreds of endemicspecies on the Hawaiian island chain; originating from a single common ancestor (supported from molecular analysis). Studies consistently find that colonization of each island occurred from older to younger islands, speciating peripatrically at least fifty percent of the time. In conjunction with *Drosophila*, Hawaiian lobeliads (*Cyanea*) have also undergone an adaptive radiation, with upwards of twenty-seven percent of extant species arising after new island colonization—exemplifying peripatric speciation—once again, occurring in the old-to-young direction.

Several other endemic species in Hawaii also provide evidence of peripatric speciation such as the endemic flightless crickets (*Laupala*). Shaw estimated that, "17 species out of 36 well-studied cases of speciation were peripatric". Gillespie and Croom also found evidence of peripatry in Hawaiian *Tetragnatha* spiders.

Coyne and Orr contend that evidence of peripatry arising on continents is far more difficult to detect due to the possibility of vicarient explanations being equally likely. However, studies concerning the Californian plant species *Clarkia biloba* and *C. biloba* strongly suggest a peripatric origin. In addition, a great deal of research had been conducted on several species of land snails involving chirality that suggests peripatry.

Sympatric Speciation

It does not require large area to reduce gene flow between parts of a population

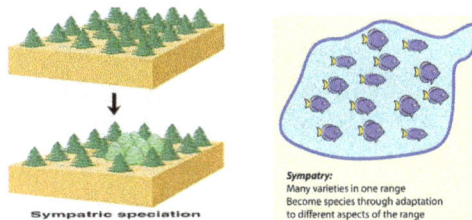

Sympatry:
Many varieties in one range
Become species through adaptation
to different aspects of the range

Sympatric speciation

Sympatric speciation

Sympatric speciation is speciation that occurs when two groups of the same species live in the same geographic location, but they evolve differently until they can no longer interbreed and are considered different species. It is different from other types of speciation, which involve the formation of a new species when a population is split into groups via a geographic barrier or migration. Sympatric speciation can be seen in many different types of organisms including bacteria, cichlid fish, and the apple maggot fly, but it can be difficult to tell when sympatric speciation is occurring or has occurred in nature.

Examples of Sympatric Speciation

In Bacteria

True examples of sympatric speciation have rarely been observed in nature. Sympatric speciation is thought to occur more often in bacteria, because bacteria can exchange genes with other individuals that aren't parent and offspring in a process known as horizontal gene transfer. Sympatric speciation has been observed in *Bacillus* and *Synechococcus* species of bacteria, and in the bacterioplankton *Vibrio splendidus*, among others. Subgroups of species that are undergoing sympatric speciation will show few differences since they have been diverging for a relatively recent time on the slow timescale at which evolution takes place. It is thought that one important factor in cases of sympatric speciation is adaptation to environmental conditions; if some members are specialized for living in a certain environment, that subgroup may go on to occupy a different environmental niche and eventually evolve into a new species over time.

In Cichlids

This is a photo of a Midas cichlid.

Another example of sympatric speciation is found in two species of Midas cichlid fish (*Amphilophus* species), which live in Lake Apoyo, a volcanic crater lake in Nicaragua. Researchers analyzed the DNA, appearance, and ecology of these two closely related species. The two species, though overall very similar, do have slight differences in appearance, and they cannot interbreed. All available evidence suggests that one species evolved from the other, which is the species of Midas cichlids that originally colonized the lake. The newer species evolved relatively recently, but in evolutionary terms, this means that it is thought to have evolved less than 10,000 years ago.

In Apple Maggot Flies

An extremely recent example of sympatric speciation may be occurring in the apple maggot fly, *Rhagoletis pomonella*. Apple maggot flies used to lay their eggs only on the fruit of hawthorn trees, but less than 200 years ago, some apple maggot flies began to lay their eggs on apples instead. Now there are two groups of apple maggot flies: one that lays eggs on hawthorns and one that lays eggs on apples. Males look for mates on the same type of fruit that they grew on, and females lay their eggs on the same type of fruit that they grew up on. Therefore, flies that grew up on hawthorns will raise offspring on hawthorns, and flies that grew up on apples will raise offspring on apples. There are already genetic differences between the two groups, and over a long period of time, they could become separate species. This shows how speciation can occur even when different subgroups of the same species have the same geographic range.

Cospeciation

Cospeciation is the process whereby one population speciates in response to and in concert with another, and is a consequence of the associates dependence on its host for its survival.

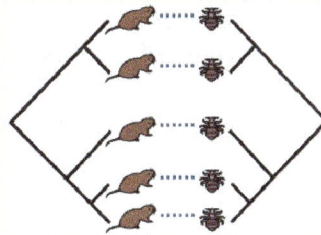

Cospeciation

- parallel host and parasite phylogenies is evidence of cospeciation.
- This example is somewhat idealized—rarely do scientists find hosts and parasites with exactly matching phylogenies.
- However, sometimes the phylogenies indicate that cospeciation did happen along with some host-switching.

The fact that some groups of parasites are closely adapted to a particular host species or small group of host species means that transmission from one host species to another is relatively rare in these parasites even on an evolutionary scale or occurs only within a group of related species. However, this means that, as the host species is subject to speciation, i.e. individual daughter species split off during evolution, the parasite must also undergo speciation. Cospeciation occurs in the parasite and the host. En-forced speciation of the parasite may be caused by *genetic drift* following separation of a subpopulation bound to both daughter species and thus after separation of their gene pools. In many cases, however, this is apparently also strengthened by *natural selection*, adaptation to two, now mutually different, host species. Of course, some-times host speciation need not necessarily be followed by speciation of the parasite; a single species of parasite can attack a relatively broad group of mutually related spe-cies. However, such a situation generally occurs when the host species live in the same territory (sympatrically) or when the parasite is spread by intermediate host species, which have sufficient mobility and habitats encompassing the habitats of all the related host species.

The fact that the sequence of parasite speciation must frequently copy or at least respect the sequence of speciation of the host taxon means that the *cladograms* (*phylogenet-ic trees*) of the host and parasite will be similar to a certain degree. Thus, if we know the *phylogenetic tree* of the host taxon, it is often possible to approximately estimate the *phylogenetic trees* of its parasites (and vice versa). This Fahrenholz rule was first apparently published in 1896 by Kellogg. It is, of course, possible that, in some cases, the parasite can transfer to some other host organism, especially if the parasite is not particularly specialized, i.e. has a broader host spectrum. A parasite can also "miss the boat", i.e. may be absent at the critical moment in the host subpopulation that, in time, results in the formation of an independent species, or could become extinct over time in a host species. In this case, the *phylogenetic trees* of the parasite and host need not be identical or can contain individual anomalies that are externally manifested, for exam-ple by crossing of the relevant branches of the cladograms. Empirical data indicate that better similarity is exhibited between the cladograms of hosts and their parasites when the relationship of the two organisms tends to be mutualistic and that the similarity is less in cases of actual parasitism. For example, the cladograms of *Wolbachia* bacteria in insects are far less similar to the cladograms of their hosts than those of *Wolbachia* bacteria and *Filaria*. This is apparently a result of the fact that a coevolutionary "arms race" occurs between the parasite and the host, in which the host occasionally manages to successfully shake off its parasite. However, this success is usually only temporary; the empty niche can be secondarily occupied by a different species of parasite that is often not related to the original species.

Despeciation

Despeciation refers to an evolutionary process whereby a given animal species is lost due to the combination of another species. Therefore, Despeciation is the oppo-site of speciation which entails the emergence of a new kind of species. The process is uncommon and is almost the same as extinction because of the loss of given species.

Thus, in a population, a unique species is absorbed by another one hence the end of its lineage occurs.

Despeciation in Ravens

A study has revealed that two raven species may soon merge into one. As a result, the research contradicted earlier studies which showed splitting of a given species into two due to speciation. The analysis, courtesy of Washington, suggested that the hybridization of two divergent raven species may be interbred giving rise to a single species.

Initially, the Common Ravens were treated as one. However, in the year 2000, it was discovered that the ravens consisted of two main lineages. These lineages are "California," which is found mostly to the southwestern regions of the United States, and the "Holarctic," found in other regions such as Alaska, Maine (both in the United States), Russia, and Norway, among other areas. Thus, the report unmasked the multiple raven species which had appeared as one over many years.

Extensive research on the raven DNA was conducted in the lab in Omland. The results showed that despite the ravens consisting of two lineages, both are widely mingled. As per the research, the two raven lineages diverged from a common ancestor millions of years ago. Because of the two lineages coming together, cases of hybridization between them have taken place over recent decades. Subsequently, there is an emergence of a unique species which is a combination of the two.

Therefore, despeciation has led to the speciation reversal. In this case of ravens, two lineages are traced back to one common origin, but evolution has led to each lineage developing new characteristics. When they later come together and interbreed, they act as if they are going back to their initial state of sharing the common ancestor. Hence, the term, "speciation reversal" because the lineages are in a converging.

Merits and Demerits of Despeciation

The merging of two lineages of a given species may have a significant impact on the population of the organism. The species is being reversed back into its old characteristics. Therefore, there might be adverse effects which may have made their ancestors diverge to cope with the ever-changing environmental conditions. Subsequently, the species may be subjected to extinction just like their predecessors.

Other Animals that have Undergone Despeciation

Apart from ravens, some other species also exhibit despeciation. For example, human beings are considered to have undergone despeciation due to the genomes associated with early human remains like the Denisovans and Neanderthals, among others. Also, studies carried out by Tylor about the three-spined sticklebacks in the southwestern region of the British Columbia gave exciting results about despeciation.

The relationship between species loss and ecosystem function has been an area of intense focus in ecology, with many studies highlighting the impact of extinction on community and ecosystem processes. Yet studies to date have focused exclusively on the effects of extinction by demographic decline, and little is known about how the alternative extinction process, reverse speciation through introgressive hybridization, impacts ecosystems. Species communities that have assembled by recent adaptive radiation in otherwise species-poor environments seem especially prone to introgressive extinction, in part because reproductive barriers may be highly sensitive to environmental context. These species are often young yet highly differentiated ecologically, showing strong divergence in both resource use and functional traits, such as gill raker number in zooplanktivorous fishes and beak size in seed-eating finches. Here we show that reverse speciation can have wide-ranging ecological effects. These impacts were largely on prey species and were found to extend out from modifications of the abundance of aquatic insects to aquatic ecosystem function and even to the aerial environment through modifications of the abundance of emerging insects.

Studies to date that have documented reverse speciation find that demographic decline may occur simultaneously, leading to genetic idiosyncrasies that shape the morphology of the resulting population. Hybridization during introgressive extinction most likely also produces a wide range of novel phenotype combinations, which could be influenced by mate choice asymmetries and might be subject to strong natural selection Our study suggests that these case-specific idiosyncrasies may be important in ultimately determining the ecological impacts of introgressive extinction. Nevertheless, our data demonstrate that the phenotypic changes associated with reverse speciation can provide some predictability to the ecological effects of this rapid phenotypic shift. Utilizing this relationship between morphology and ecological impacts could lend some predictive power when projecting the impacts of evolutionary changes on ecosystems.

The number of cases of reverse speciation is growing, and with most examples to date invoking anthropogenic habitat alterations as a driving factor they seem likely to continue to climb.

Hybrid Speciation

'Hybrid speciation' implies that hybridization has had a principal role in the origin of a new species. The definition applies cleanly to hybrid species that have doubled their chromosome number (allopolyploidy): derived species initially contain exactly one genome from each parent, a 50% contribution from each, although, in older polyploids, recombination and gene conversion may eventually lead to unequal contributions. Furthermore, allopolyploids are largely reproductively isolated by ploidy. Recombinational hybrid speciation, in which the genome remains diploid (homoploid hybrid speciation), is harder to define. The fraction from each parent will rarely be 50% if backcrossing is involved. Homoploid hybrid species may be only weakly reproductively isolated, and are hard to distinguish from species that gain alleles by hybridization and introgression, or from persistent ancestral polymorphisms.

Hybrid speciation is only possible if reproductive isolation is weak; if hybrids are intermediate, hybrid species will be even more weakly isolated.

A hybrid species will then be a third cluster of genotypes, a hybrid form that has become stabilized and remains distinct when in contact with either parent.

Hybridization can also influence speciation by means of 'reinforcement', where mating barriers evolve owing to selection against unfit hybrids. Although hybridization contributes to speciation, I do not consider reinforcement to be hybrid speciation, because a third species does not form. A related and highly relevant phenomenon is 'hybridogenesis'. The diploid or triploid edible frog Rana esculenta is a well known example: it is heterozygous for complete Rana lessonae and Rana ridibunda genomes.

Theory and Background of Hybrid Speciation

Hybridization may be "the grossest blunder in sexual preference which we can conceive of an animal making", but it is nonetheless a regular event. The fraction of species that hybridize is variable, but on average around 10% of animal and 25% of plant species are known to hybridize with at least one other species. Hybridization is especially prevalent in rapidly radiating groups: 75% of British ducks (Anatidae), for example. Recent, closely related species are most likely to hybridize, although hybridization and introgression may often persist for millions of years after initial divergence. Hybridization is thus a normal feature of species biology, if at a rather extreme end of the natural spectrum of sexuality - it is not merely an unnatural "breakdown of isolating mechanisms". At the population level, interspecific hybrids are, of course, unusual, forming, 0.1% of individuals in a typical population; they are also 'hopeful monsters', with hefty differences from each parent, no adaptive history to any ecological niche, and little apparent scope for survival. Furthermore, hybrids are often sterile or in viable owing to divergent evolution in each species. Even if a healthy hybrid is formed, it normally suffers 'minority cytotype disadvantage' because it encounters few mates of its own type, and backcrosses to the more abundant parent species will often be unfit. For example, a rare tetraploid hybrid will produce unfit triploid progeny with diploid parents.

Yet hybrid species exist. What advantages could outweigh the catalogue of difficulties? This innocent-sounding question plunges to the heart of controversies about adaptive evolution. Is saltational evolution possible? Are maladaptive intermediates and genetic drift involved? Common sense and prevailing opinion suggests that evolution normally occurs by small adjustments rather than saltation, and rarely involves maladaptation. It is therefore extraordinary that hybrid speciation can disobey both rules. Hybridization (or hybridogenesis) can act as a multi-locus 'macro-mutation' that reaches out over large phenotypic distances5 to colonize unoccupied ecological niches or adaptive peaks. Furthermore, random drift in small, localized hybrid populations provides a parsimonious solution to maladaptation, to enable local establishment, stabilization and ultimate spread.

Two principal types of hybrid speciation are treated here: allopolyploidy and homoploid hybrid speciation.

Hybrid Speciation Through Allopolyploidy

Polyploidy is a well-established speciation mode in plants, although many aspects of polyploid evolution are only today being revealed. Speciation can be via autopolyploidy (duplication of chromosomes within a species) or allopolyploidy (duplication of chromosomes in hybrids between species), although the boundary is blurred because of the 'fuzzy' nature of species. Polyploid species are reproductively isolated from their parents because when polyploids mate with diploids, progeny with odd-numbered ploidies, such as triploids, are produced. These offspring may be viable but typically produce sterile gametes with unbalanced chromosomal complements (aneuploidy). Polyploidy is thus a simple saltational means of achieving speciation. The process may be repeated many times, leading to lineages with.80-fold ploidy in some vascular plants - 40–70% of all plant species are polyploids.

Allopolyploid speciation can result from somatic chromosome doubling in a diploid hybrid, followed by selfing to produce a tetraploid. This was the route taken by Primula kewensis, the allopolyploid that arose spontaneously in 1909 among cultivated diploid hybrids of Primula verticillata and Primula floribunda. However, there are other possibilities, such as fusion of two unreduced gametes after failure of reduction divisions in meiosis. A third route is the 'triploid bridge', in which rare, unreduced (diploid) gametes fuse with normal haploid gametes to form triploids. Triploids are normally sterile, but can contribute to tetraploid formation by themselves producing occasional, unreduced triploid gametes that can backcross with a normal haploid gamete to form tetraploid progeny. This was the route used to engineer the first, and maybe the only, synthetic (that is, in the laboratory), self-sustaining bisexual animal polyploid strain, a hybrid between silk moths (Bombyx mori and Bombyx mandarina).

The hyperspace of possible phenotypes and genotypes can be represented as an adaptive landscape. Fitness optima ('adaptive peaks') are coloured blue. Adaptive landscapes are not rigid, but are readily distorted by environmental or biotic changes, including evolutionary change. Mean phenotypes of species and their hybrids are shown as crosses, and offspring distributions as dots. Species 1 and 2 are adapted to different fitness optima. Natural selection acts mainly within each species, so hybrids are 'hopeful monsters', far from phenotypic optima (solid arrows). It is therefore hard to imagine how hybrids often attain new optima unless unoccupied adaptive peaks are abundant. Polyploid hybrids can have a variety of advantages over their parents, including heterozygote advantage, extreme phenotypic traits and reproductive isolation. Genetic variation in their offspring will initially be similar to that of non-hybrid parents if recombination between parental genomes is rare; such hybrids will not spread unless already near an optimum. Homoploid hybrids have fewer initial advantages, but their progeny can have extremely high genetic variances via recombination, including phenotypes more extreme than either parent—transgressive variation. This burst of variation can help homoploids attain new adaptive peaks (dotted arrow) far from parental optima.

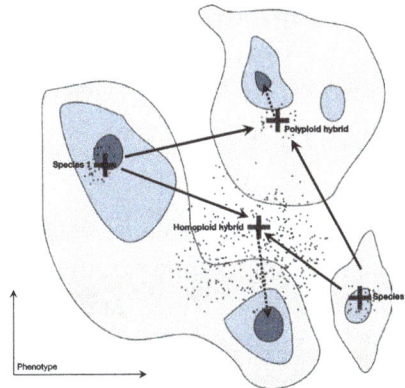

Figure: Hybridization and the adaptive landscape

After polyploid hybrids arise, they still face major hurdles. Diploid and triploid hybrids are strongly disfavoured because their aneuploid gametes are almost always sterile. Even when even-numbered allopolyploidy is achieved, chromosome pairing is rarely perfect. Furthermore, assuming new polyploids are rare, they will mate mostly with incompatible parentals, leading to minority cytotype disadvantage19. These problems almost certainly explain why bisexual polyploid speciation is more common in plants than animals:

(1) Plants usually have indeterminate growth, and somatic chromosome doubling can lead to germline polyploidy;

(2) Plants are also often perennial or temporarily clonal, allowing multigenerational persistence of hybrid cell lines within which polyploid mutations can occur;

(3) Plants are more often hermaphrodites, allowing selfing as a means of sexual reproduction of rare polyploids, once formed;

(4) Gene flow is weaker in plants than in animals, and local populations with unusual ploidy (whether by local drift or selection) can form more readily to overcome minority cytotype disadvantage.

As expected, polyploidy is strongly associated with asexual reproduction, selfing and perenniality in plants, as well as in animals. In clonal polyploid animals, reversion to out-crossing is rare, whereas in plants, with frequent alternation between clonal and sexual phases, bisexual polyploid species are common and themselves often give rise to further species. Thus, animal allopolyploids such as stick insects (Bacillus) and fresh-water snails (Bulinus truncatus) are often, although not always, parthenogenetic or selfers. Muller's theory that sex chromosomes in animals prevent sexual polyploidy owing to sex:autosome gene dosage is no longer given much credence.

How common is polyploid speciation? Otto and Whitton provided new insights from the over-representation of even-numbered chromosome counts. Recent polyploidy explains, ~ 2 − 4% of speciation events in flowering plants and, ~ 7% of speciation events in ferns, and these are probably underestimates6 (40−70% of plant species overall are polyploid, but this includes the effects of much non-polyploid speciation within already polyploid lineages). In animals, there is no bias towards even-numbered chromosomal counts, suggesting that animal polyploid speciation is very rare compared with other speciation modes.

Traditional dogma has it that allopolyploids arise more readily than autopolyploids because the latter are more prone to chromosome pairing problems in meiosi ; however, this view is no longer generally accepted. Newly arisen autopolyploids have levels of infertility and aneuploid gametes comparable to those of allopolyploids. Furthermore, many autopolyploids probably lie unrecognized by taxonomy within diploid progenitor species. Yet these discoveries give little insight into a more important question: what fraction of polyploids that spread successfully are allopolyploids? Autopolyploids may often be doomed to extinction, perhaps through competition with similar diploid relatives. Opinions differ, but it probably remains true that allopolyploids are more successful than autopolyploids; certainly allopolyploids are a sizeable fraction of wellstudied crop cases, such as wheat, cotton and tobacco. There are almost no surveys of entire floras, although a small-scale survey in the United States revealed that 79−96% of 28 polyploid species were allopolyploids. Recently, the Arctic flora was surveyed, in which about 50% of the often clonal or selfing species are polyploids. In Svalbard (Spitsbergen), 78% of the 161 species are polyploid, with the average level of ploidy approximately hexaploid. Every one of the 47 polyploid species studied genetically shows fixed marker heterozygosity, implying 100% allopolyploidy. The Arctic is, of course, an extreme environment, but this remains the most comprehensive survey so far. If Svalbard is typical, most successful polyploids are also hybrid species.

After formation, novel allopolyploids face the usual 'hopeful monster' difficulties. It helps if they can exploit a new ecological niche that is both vacant and also spatially separated to ameliorate minority cytotype disadvantage. For example, recent allopolyploid hybrids between introduced and native plants have successfully spread from sites of origin (for example, Senecio cambrensis in Wales and Spartina anglica in England). These invasive allopolyploids were able to exploit vacant ecological roles with relatively little evolutionary change.

Stochastic drift may also be necessary to overcome minority cytotype and other disadvantages. A few polyploids, usually from the same hybridization event, must accumulate locally for the process to take off, probably involving chance or an unusual local selective regime. Stochastic effects are evident in nature. An independently derived Scottish population of S. cambrensis became extinct in Edinburgh some 20 yr after being discovered. Of two origins, only the Welsh population now survives. Other allopolyploid hybrids can arise repeatedly from the same parents, but many widespread polyploids (for example, S. anglica) probably originated only once or a few times even though parent species are in broad contact, again showing the importance of chance in the origins of hybrid species.

Recombinational and Homoploid Hybrid Speciation

Homoploid hybrid speciation or recombinational speciation is wellknown in flowering plants. Speciation takes place in sympatry (by definition, as hybridization requires gene flow). Hybrids must then overcome chromosome and gene incompatibilities, while lacking reproductive isolation via polyploidy. For these reasons, the process is often considered unlikely.

However, hybridization can boost genetic variance, allowing colonization of unexploited niches. Suppose + and − alleles at genes affecting a quantitative trait differ between species, so that each has fixed differences $(+++--$ and $---++,$ say$)$. Recombination can then liberate 'transgressive' quantitative variation, often more extreme than either parent (for example, $-----$ and $+++++$). Most early recombinants will be unfit, but extreme hybrids can colonize niches unavailable to parents. If ecological opportunities are partially separated from the parental habitat, if like hybrids tend to associate (for example, by means of seasonality or drift in small populations), or if selfing or inbreeding is common, gene flow between hybrids and parents will be reduced and hybrid speciation becomes more likely. Successful hybrid species might also displace one or more parent species ecologically, and obliterate evidence of their own hybrid origin.

In plants, about 20 well-established homoploid hybrid species are known, but they are hard to detect and may be more prevalent. The best documented are the desert sunflowers Helianthus anomalus, Helianthus deserticola and Helianthus paradoxus, which all derive from hybrids between mesic-adapted Helianthus annuus and Helianthus petiolaris. Selfing is rare and provides little assistance to establishment, but the three hybrid species survive drought better than their parents, suggesting recruitment of hybrid transgressive variation. Synthetic hybrid populations are readily recreated with karyotypic combinations like those in wild hybrid species, because selection repeatedly favours similar combinations of compatible chromosomal rearrangements. In addition, the wild contribution from each parent of extreme adaptive traits for morphology, physiology and life history of the hybrids (for example, small leaf size, seed dormancy, or tolerance of drought and salt) matches experimental predictions. In Helianthus, recombinant genotypes and spatial separation have enabled the hybrids to flourish where their parents are absent.

Although bisexual polyploids are often barred in animals, there is no reason why homoploid hybrid species would be rarer in animals than in plants. The number of cases in animals is growing rapidly. A recent example is the invasive sculpin, a hybrid fish derived from the Cottus gobio group from the Scheldt River (compare Cottus perifretum) and upper tributaries of the Rhine (compare Cottus rhenanum). Sculpins are normally restricted to clear, well-oxygenated cold waters in upper river tributaries across Europe. The Rhine and Scheldt rivers became connected as a result of earlier canal building, but invasive sculpins appeared in the warmer and muddier lower Rhine only in the past fifteen years. Morphologically, the invasive sculpin is intermediate, and its mitochondrial DNA, as well as nuclear single nucleotide polymorphisms and microsatellites, are characteristic of both Scheldt and Rhine forms. The hybrid form meets upper Rhine sculpins in narrow hybrid zones, but remains distinct despite gene flow, suggesting that it is adapted mainly to the lower Rhine. Recent evolution and spread of the invasive sculpin, as well as intermediacy, provides convincing evidence of adaptive hybrid origin. A more ancient example is the cyprinid fishGila seminuda, which inhabits the Virgin River, a tributary of the Colorado River (USA). This hybrid species contacts but does not overlap its parent species, Gila robusta and Gila elegans, from the Colorado River. Gila seminuda is morphologically and genetically intermediate, and similar to synthetic hybrids. Intermediacy may allow it to out-compete both parents in the Virgin River35. A similar case, with well-documented genetic intermediacy, is an unnamed form of the butterfly genus Lycaeides. This homoploid hybrid species uses a different host plant and inhabits high-elevation alpine habitats unoccupied by either parent.

Rhagoletis fruitflies provide another historically documented example. In 1997 flies were first found on introduced honeysuckle. Molecular markers in the fly are a blend from the two parents: the blueberry maggot Rhagoletis mendax and the snowberry maggot Rhagoletis zephyria. No F1 genotypes are detected between the hybrid form and its parents where they overlap; the new fly is reproductively isolated. Rhagoletis flies mate on host fruits, and host choice ensures mating specificity. A different route to hybrid speciation can be inferred from the Colombian butterfly Heliconius heurippa. This form has a colour pattern like that of synthetic hybrids between local Heliconius cydno and Heliconius melpomene. Microsatellite alleles are shared across all three species, but H. heurippa forms an allele frequency cluster distinguishable from either parent. The hybrid wing coloration of H. heurippa is a cue in mating discrimination, and directly causes reproductive isolation from both parents. Similarly, in the fish Xiphophorus clemenciae, the 'swordtail', a hybridization-derived trait, is involved in sexual selection and mate choice and may be related to its speciation.

Hybrid speciation in animals is supported so far only by lowresolution molecular data. Genomic mapping of ecological or speciation-related hybrid traits, which so strongly supports hybrid speciation in Helianthus, is not yet available for any animal case. Many homoploid hybrid species fail to overlap with at least one parent species, and reproductive isolation is weak, so species status could be questioned (the Lonicera-feeding Rhagoletis is an exception). Nonetheless, in the cases surveyed here, hybrid traits often contribute strongly to maintenance or ecological expansion of the new form.

Is hybrid speciation important in evolution?

There are now many examples of hybrid species. We know that polyploidy is common in plants, giving rise to $\geq 2-7\%$ of vascular plant species, but rarer in animals. Furthermore, ancient polyploidy has been found at the root of many plant and animal groups. Genome duplications probably facilitated the evolution of complex organisms (although this is debated), and we can infer that successful genome duplications were mostly allopolyploid, provided that limited plant community data are reliable. Hybridization would then be a catalyst not only for speciation but also for major evolutionary innovations.

Polyploid speciation leaves a clear genomic signature, but we have little idea how common homoploid hybrid species are. They could be abundant: most speciation involves natural selection; natural selection requires genetic variation; genetic variation is enhanced by hybridization; and hybridization and introgression between species is a regular occurrence, especially in rapidly radiating groups. Enough suspected animal homoploid hybrid species exist to indicate that it may be at least as common as in plants, in contrast to the situation for polyploidy, where a variety of traits prevent its occurrence. It now seems intuitively unlikely that all biodiversity arose as a result of recombination of existing diversity, but homoploid hybrid species might still represent a large fraction. Nonetheless, there are few convincing cases, probably, in part, because of the difficulty of demonstrating that hybridization has led to speciation. We clearly need more genomic analyses. As for hybrid species as a whole, we have observed recent speciation in the laboratory or nature in seven genera discussed here (Helianthus, Senecio, Primula, Spartina, Rhagoletis, Bombyx and Cottus), and there are many other cases. It would be hard to find another mode of speciation so readily documented historically and so amenable to experimentation.

That hybrid species exist at all reveals something perhaps unexpected about adaptive landscapes. If hybrid 'hopeful monsters', with all their problems, are ever to survive in competition with their parents, they must be able to hit (and for polyploid species, hit almost exactly) new adaptive combinations of genes. This implies both that many adaptive peaks are scattered about in the adaptive landscape, and also that many are unoccupied. Liberal adaptive landscapes are further supported by the successes of many introduced species, and by fossil evidence: for insects, angiosperms and many other groups, diversity seems to have been increasing more or less continuously over geological time.

The ability of hybrid species to invade hitherto unoccupied niches also means that hybridization can contribute to adaptive radiations such as African cichlid fish and

Darwin's finches. This principle is well demonstrated by the 'domestication niche'. Humans have unwittingly created many allopolyploid and other hybrid crops and domestic animals while selecting for transgressively high yields. Even our own species may have a hybrid genomic ancestry, although this is contested. Whichever way the debate about humans is resolved, it would be hardly surprising if hybridization was one trigger for the origin of Homo sapiens, the most invasive mammal on the planet.

References

- Ridley, Mark. "Speciation - What is the role of reinforcement in speciation?". Retrieved 2015-09-07. Adapted from Evolution (2004), 3rd edition (Malden, MA: Blackwell Publishing), ISBN 978-1-4051-0345-9

- Maynard Smith, John (December 1983). «The Genetics of Stasis and Punctuation». Annual Review of Genetics. 17: 11–25. doi:10.1146/annurev.ge.17.120183.000303. PMID 6364957

- Ollerton, Jeff (September 2005). "Speciation: Flowering time and the Wallace Effect"(PDF). Heridity. 95 (3): 181–182. doi:10.1038/sj.hdy.6800718. PMID 16077739. Archived from the original (PDF) on 2007-06-05. Retrieved 2015-09-07

- Tarkhnishvili, David; Murtskhvaladze, Marine; Gavashelishvili, Alexander (August 2013). "Speciation in Caucasian lizards: climatic dissimilarity of the habitats is more important than isolation time". Biological Journal of the Linnean Society. 109 (4): 876–892. doi:10.1111/bij.12092

- Speciation, encyclopedia: nationalgeographic.org, Retrieved 15 July 2018

- Gould, Stephen Jay (1980). A Quahog is a Quahog. The Panda's thumb. More reflections in natural history. New York: W. W. Norton & Company. pp. 204–213. ISBN 0-393-30023-4

- Lawson, Lucinda P.; Bates, John M.; Menegon, Michele; Loader, Simon P. (2015). "Divergence at the edges: peripatric isolation in the montane spiny throated reed frog complex". BMC Evolutionary Biology. 15 (128). doi:10.1186/s12862-015-0384-3

- Sympatric-speciation: biologydictionary.net, Retrieved 20 June 2018

- Minkel, J. R. (September 8, 2006). "Wandering Fly Gene Supports New Model of Speciation". Scientific American. Retrieved 2015-09-11

- Koeslag, Johan H. (December 21, 1995). "On the Engine of Speciation". Journal of Theoretical Biology. 177 (4): 401–409. doi:10.1006/jtbi.1995.0256. ISSN 0022-5193

- What-is-despeciation: worldatlas.com, Retrieved 31 March 2018

Chapter 5

Mutation and Adaptation

A permanent alteration in the nucleotide sequence of a genome of an organism is called mutation. Adaptation is a dynamic evolutionary process, which fits organisms into their surroundings, and enhances their evolutionary fitness. All the diverse principles of mutation and adaptation have been carefully analyzed in this chapter, such as mutationism, neutral mutations, suppressor mutation, point mutation, dynamic mutation, frameshift mutation, etc.

Mutation

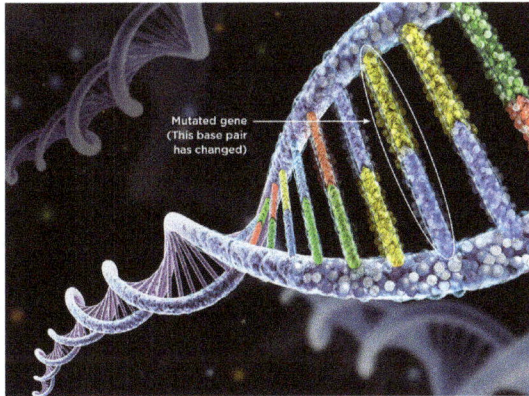

Mutation is an alteration in the genetic material (the genome) of a cell of a living organism or of a virus that is more or less permanent and that can be transmitted to the cell's or the virus's descendants. (The genomes of organisms are all composed of DNA, whereas viral genomes can be of DNA or RNA; Mutation in the DNA of a body cell of a multicellular organism (somatic mutation) may be transmitted to descendant cells by DNA replication and hence result in a sector or patch of cells having abnormal function, an example being cancer. Mutations in egg or sperm cells (germinal mutations) may result in an individual offspring all of whose cells carry the mutation, which often confers some serious malfunction, as in the case of a human genetic disease such as cystic fibrosis. Mutations result either from accidents during the normal chemical transactions of DNA, often during replication, or from exposure to high-energy electromagnetic radiation (e.g., ultraviolet light or X-rays) or particle radiation or to highly reactive chemicals in the environment. Because mutations are random changes, they are expected to be mostly deleterious, but some may be beneficial in certain environments.

In general, mutation is the main source of genetic variation, which is the raw material for evolution by natural selection.

The genome is composed of one to several long molecules of DNA, and mutation can occur potentially anywhere on these molecules at any time. The most serious changes take place in the functional units of DNA, the genes. A mutated form of a gene is called a mutant allele. A gene is typically composed of a regulatory region, which is responsible for turning the gene's transcription on and off at the appropriate times during development, and a coding region, which carries the genetic code for the structure of a functional molecule, generally a protein. A protein is a chain of usually several hundred amino acids. Cells make 20 common amino acids, and it is the unique number and sequence of these that give a protein its specific function. Each amino acid is encoded by a unique sequence, or codon, of three of the four possible base pairs in the DNA (A–T, T–A, G–C, and C–G, the individual letters referring to the four nitrogenous bases adenine, thymine, guanine, and cytosine). Hence, a mutation that changes DNA sequence can change amino acid sequence and in this way potentially reduce or inactivate a protein's function. A change in the DNA sequence of a gene's regulatory region can adversely affect the timing and availability of the gene's protein and also lead to serious cellular malfunction. On the other hand, many mutations are silent, showing no obvious effect at the functional level. Some silent mutations are in the DNA between genes, or they are of a type that results in no significant amino acid changes.

Mutations are of several types. Changes within genes are called point mutations. The simplest kinds are changes to single base pairs, called base-pair substitutions. Many of these substitute an incorrect amino acid in the corresponding position in the encoded protein, and of these a large proportion result in altered protein function. Some base-pair substitutions produce a stop codon. Normally, when a stop codon occurs at the end of a gene, it stops protein synthesis, but, when it occurs in an abnormal position, it can result in a truncated and nonfunctional protein. Another type of simple change, the deletion or insertion of single base pairs, generally has a profound effect on the protein because the protein's synthesis, which is carried out by the reading of triplet codons in a linear fashion from one end of the gene to the other, is thrown off. This change leads to a frame shift in reading the gene such that all amino acids are incorrect from the mutation onward. More-complex combinations of base substitutions, insertions, and deletions can also be observed in some mutant genes.

Mutations that span more than one gene are called chromosomal mutations because they affect the structure, function, and inheritance of whole DNA molecules (microscopically visible in a coiled state as chromosomes). Often these chromosome mutations result from one or more coincident breaks in the DNA molecules of the genome (possibly from exposure to energetic radiation), followed in some cases by faulty rejoining. Some outcomes are large-scale deletions, duplications, inversions, and translocations. In a diploid species (a species, such as human beings, that has a double set of chromosomes in the nucleus of each cell), deletions and duplications alter gene balance and often result in abnormality. Inversions and translocations involve no loss or gain and are functionally normal unless a break occurs within a gene. However, at meiosis (the specialized nuclear divisions that take place during the production of gametes—i.e., eggs and sperm), faulty pairing of an inverted or translocated chromosome set with a normal set can result in gametes and hence progeny with duplications and deletions.

Loss or gain of whole chromosomes results in a condition called aneuploidy. One familiar result of aneuploidy is Down syndrome, a chromosomal disorder in which humans are born with an extra chromosome 21 (and hence bear three copies of that chromosome instead of the usual two). Another type of chromosome mutation is the gain or loss of whole chromosome sets. Gain of sets results in polyploidy—that is, the presence of three, four, or more chromosome sets instead of the usual two. Polyploidy has been a significant force in the evolution of new species of plants and animals.

Most genomes contain mobile DNA elements that move from one location to another. The movement of these elements can cause mutation, either because the element arrives in some crucial location, such as within a gene, or because it promotes large-scale chromosome mutations via recombination between pairs of mobile elements in different locations.

At the level of whole populations of organisms, mutation can be viewed as a constantly dripping faucet introducing mutant alleles into the population, a concept described as mutational pressure. The rate of mutation differs for different genes and organisms. In RNA viruses, such as the human immunodeficiency virus (HIV), replication of the genome takes place within the host cell using a mechanism that is prone to error. Hence, mutation rates in such viruses are high. In general, however, the fate of individual mutant alleles is never certain. Most are eliminated by chance. In some cases a mutant allele can increase in frequency by chance, and then individuals expressing the allele can be subject to selection, either positive or negative. Hence, for any one gene the frequency of a mutant allele in a population is determined by a combination of mutational pressure, selection, and chance.

Mutagens

Chemical Mutagens change the sequence of bases in a DNA gene in a number of ways:

- Mimic the correct nucleotide bases in a DNA molecule, but fail to base pair correctly during DNA replication.

- Remove parts of the nucleotide (such as the amino group on adenine), again causing improper base pairing during DNA replication.

- Add hydrocarbon groups to various nucleotides, also causing incorrect base pairing during DNA replication.

Radiation, High energy radiation from a radioactive material or from X-rays is absorbed by the atoms in water molecules surrounding the DNA. This energy is transferred to the electrons which then fly away from the atom. Left behind is a free radical, which is a highly dangerous and highly reactive molecule that attacks the DNA molecule and alters it in many ways. Radiation can also cause double strand breaks in the DNA molecule, which the cell's repair mechanisms cannot put right.

Sunlight contains ultraviolet radiation (the component that causes a suntan) which, when absorbed by the DNA causes a cross link to form between certain adjacent bases. In most normal cases the cells can repair this damage, but unrepaired dimers of this sort cause the replicating system to skip over the mistake leaving a gap, which is supposed to be filled in later.

Spontaneous mutations occur without exposure to any obvious mutagenic agent. Sometimes DNA nucleotides shift without warning to a different chemical form (know as an isomer) which in turn will form a different series of hydrogen bonds with it's partner. This leads to mistakes at the time of DNA replication.

Causes

Four classes of mutations are:

(1) Spontaneous mutations (molecular decay),

(2) Mutations due to error-prone replication bypass of naturally occurring DNA damage (also called error-prone translesion synthesis),

(3) Errors introduced during DNA repair,

(4) Induced mutations caused by mutagens.

Scientists may also deliberately introduce mutant sequences through DNA manipulation for the sake of scientific experimentation.

One 2017 study claimed that 66% of cancer-causing mutations are random, 29% are due to the environment (the studied population spanned 69 countries), and 5% are inherited.

Humans on average pass 60 new mutations to their children but fathers pass more mutations depending on their age with every year adding two new mutations to a child.

Spontaneous Mutation

Spontaneous mutations occur with non-zero probability even given a healthy, uncontaminated cell. They can be characterized by the specific change:

- Tautomerism: A base is changed by the repositioning of a hydrogen atom, altering the hydrogen bonding pattern of that base, resulting in incorrect base pairing during replication.

- Depurination: Loss of a purine base (A or G) to form an apurinic site (AP site).

- Deamination: Hydrolysis changes a normal base to an atypical base containing a keto group in place of the original amine group. Examples include C → U and A → HX (hypoxanthine), which can be corrected by DNA repair mechanisms; and 5MeC (5-methylcytosine) → T, which is less likely to be detected as a mutation because thymine is a normal DNA base.

- Slipped strand mispairing: Denaturation of the new strand from the template during replication, followed by renaturation in a different spot ("slipping"). This can lead to insertions or deletions.

- Replication slippage.

Error-prone Replication by Pass

There is increasing evidence that the majority of spontaneously arising mutations are due to error-prone replication (translesion synthesis) past DNA damage in the template strand. Naturally occurring oxidative DNA damages arise at least 10,000 times per cell per day in humans and 50,000 times or more per cell per day in rats. In mice, the majority of mutations are caused by translesion synthesis. Likewise, in yeast, Kunz found that more than 60% of the spontaneous single base pair substitutions and deletions were caused by translation synthesis.

Errors Introduced during DNA repair

Although naturally occurring double-strand breaks occur at a relatively low frequency in DNA, their repair often causes mutation. Non-homologous end joining(NHEJ) is a major pathway for repairing double-strand breaks. NHEJ involves removal of a few nucleotides to allow somewhat inaccurate alignment of the two ends for rejoining followed by addition of nucleotides to fill in gaps. As a consequence, NHEJ often introduces mutations.

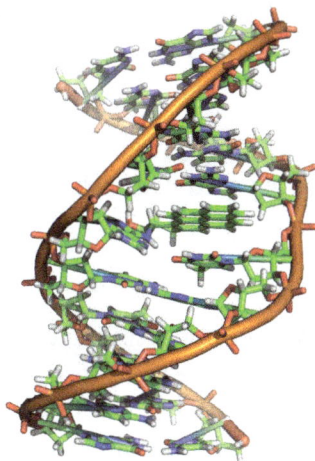

A covalent adduct between the metabolite of benzo pyrene, the major mutagen in tobacco smoke, and DNA

Induced Mutation

Induced mutations are alterations in the gene after it has come in contact with muta-gens and environmental causes.

Induced mutations on the molecular level can be caused by:

- Chemicals

  ◦ Hydroxylamine;

  ◦ Base analogs (e.g., Bromodeoxyuridine (BrdU));

  ◦ Alkylating agents (e.g., N-ethyl-N-nitrosourea (ENU)). These agents can mutate both replicating and non-replicating DNA. In contrast, a base analog can mutate the DNA only when the analog is incorporated in replicating the DNA. Each of these classes of chemical mutagens has certain effects that then lead to transitions, transversions, or deletions.

  ◦ Agents that form DNA adducts (e.g., ochratoxin A);

  ◦ DNA intercalating agents (e.g., ethidium bromide);

  ◦ DNA cross linkers;

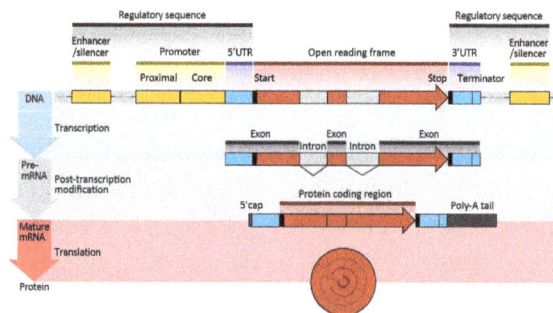

This figure depicts the following processes of transcription, splicing, and translation of a eukaryotic gene.

  ◦ Oxidative Damage;

  ◦ Nitrous acid converts amine groups on A and C to diazo groups, altering their hydrogen bonding patterns, which leads to incorrect base pairing during replication.

- Radiation

  ◦ Ultraviolet light (UV) (non-ionizing radiation). Two nucleotide bases in DNA—cytosine and thymine—are most vulnerable to radiation that can change their properties. UV light can induce adjacent pyrimidine bases in a DNA strand to become covalently joined as a pyrimidine dimer. UV radiation, in particular longer-wave UVA, can also cause oxidative damage to DNA.

○ Ionizing radiation. Exposure to ionizing radiation, such as gamma radiation, can result in mutation, possibly resulting in cancer or death.

Classification of Types

By Effect on Structure

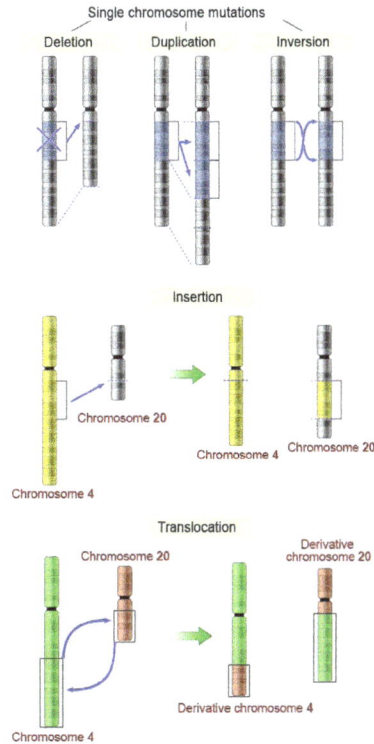

Five types of chromosomal mutations

Selection of disease-causing mutations, in a standard table of the genetic code of amino acids

The sequence of a gene can be altered in a number of ways. Gene mutations have varying effects on health depending on where they occur and whether they alter the function of essential proteins. Mutations in the structure of genes can be classified into several types.

Small-scale Mutations

Small-scale mutations affect a gene in one or a few nucleotides. (If only a single nucleotide is affected, they are called point mutations.) Small-scale mutations include:

- Insertions add one or more extra nucleotides into the DNA. They are usually caused by transposable elements, or errors during replication of repeating elements. Insertions in the coding region of a gene may alter splicing of the mRNA (splice site mutation), or cause a shift in the reading frame (frameshift), both of which can significantly alter the gene product. Insertions can be reversed by excision of the transposable element.

- Deletions remove one or more nucleotides from the DNA. Like insertions, these mutations can alter the reading frame of the gene. In general, they are irreversible: Though exactly the same sequence might in theory be restored by an insertion, transposable elements able to revert a very short deletion (say 1–2 bases) in *any* location either are highly unlikely to exist or do not exist at all.

- Substitution mutations, often caused by chemicals or malfunction of DNA replication, exchange a single nucleotide for another. These changes are classified as transitions or transversions. Most common is the transition that exchanges a purine for a purine (A ↔ G) or a pyrimidine for a pyrimidine, (C ↔ T). A transition can be caused by nitrous acid, base mispairing, or mutagenic base analogs such as BrdU. Less common is a transversion, which exchanges a purine for a pyrimidine or a pyrimidine for a purine (C/T ↔ A/G). An example of a transversion is the conversion of adenine (A) into a cytosine (C). A point mutation are modifications of single base pairs of DNA or other small base pairs within a gene. A point mutation can be reversed by another point mutation, in which the nucleotide is changed back to its original state (true reversion) or by second-site reversion (a complementary mutation elsewhere that results in regained gene functionality). As discussed below, point mutations that occur within the protein coding region of a gene may be classified as synonymous or nonsynonymous substitutions, the latter of which in turn can be divided into missense or nonsense mutations.

Large-scale Mutations

Large-scale mutations in chromosomal structure include:

- Amplifications (or gene duplications) leading to multiple copies of all chromosomal regions, increasing the dosage of the genes located within them.

- Deletions of large chromosomal regions, leading to loss of the genes within those regions.

- Mutations whose effect is to juxtapose previously separate pieces of DNA, potentially bringing together separate genes to form functionally distinct fusion genes (e.g., bcr-abl).

- Large scale changes to the structure of chromosomes called chromosomal rearrangement that can lead to a decrease of fitness but also to speciation in isolated, inbred populations. These include:

  ◦ Chromosomal translocations: interchange of genetic parts from nonhomologous chromosomes.

  ◦ Chromosomal inversions: reversing the orientation of a chromosomal segment.

  ◦ Non-homologous chromosomal crossover.

  ◦ Interstitial deletions: an intra-chromosomal deletion that removes a segment of DNA from a single chromosome, thereby apposing previously distant genes. For example, cells isolated from a human astrocytoma, a type of brain tumor, were found to have a chromosomal deletion removing sequences between the Fused in Glioblastoma (FIG) gene and the receptor tyrosine kinase (ROS), producing a fusion protein (FIG-ROS). The abnormal FIG-ROS fusion protein has constitutively active kinase activity that causes oncogenic transformation (a transformation from normal cells to cancer cells).

- Loss of heterozygosity: loss of one allele, either by a deletion or a genetic recombination event, in an organism that previously had two different alleles.

By Effect on Function

- Loss-of-function mutations, also called inactivating mutations, result in the gene product having less or no function (being partially or wholly inactivated). When the allele has a complete loss of function (null allele), it is often called an amorph or amorphic mutation in the Muller's morphs schema. Phenotypes associated with such mutations are most often recessive. Exceptions are when the organism is haploid, or when the reduced dosage of a normal gene product is not enough for a normal phenotype (this is called haploinsufficiency).

- Gain-of-function mutations, also called activating mutations, change the gene product such that its effect gets stronger (enhanced activation) or even is superseded by a different and abnormal function. When the new allele is created, a heterozygote containing the newly created allele as well as the original will express the new allele; genetically this defines the mutations as dominant phenotypes. Several of Muller's morphs correspond to gain of function, including hypermorph and neomorph. In December 2017, the U.S. government lifted a temporary ban implemented in 2014 that banned federal funding for any new "gain-of-function" experiments that enhance pathogens "such as Avian influenza, SARS and the Middle East Respiratory Syndrome or MERS viruses."

- Dominant negative mutations (also called antimorphic mutations) have an altered gene product that acts antagonistically to the wild-type allele. These mutations usually result in an altered molecular function (often inactive) and are characterized by a dominant or semi-dominant phenotype. In humans, dominant negative mutations have been implicated in cancer (e.g., mutations in genes p53, ATM, CEBPA and PPARgamma). Marfan syndrome is caused by mutations in the *FBN1* gene, located on chromosome 15, which encodes fibrillin-1, a glycoprotein component of the extracellular matrix. Marfan syndrome is also an example of dominant negative mutation and haploinsufficiency.

- Hypomorphs, after Mullerian classification, are characterized by altered gene products that acts with decreased gene expression compared to the wild type allele.

- Neomorphs are characterized by the control of new protein product synthesis.

- Lethal mutations are mutations that lead to the death of the organisms that carry the mutations.

- A back mutation or reversion is a point mutation that restores the original sequence and hence the original phenotype.

By Effect on Fitness

In applied genetics, it is usual to speak of mutations as either harmful or beneficial:

- A harmful, or deleterious, mutation decreases the fitness of the organism.

- A beneficial or advantageous mutation increases the fitness of the organism. Mutations that promote traits that are desirable are also called beneficial. In theoretical population genetics, it is more usual to speak of mutations as deleterious or advantageous than harmful or beneficial.

- A neutral mutation has no harmful or beneficial effect on the organism. Such mutations occur at a steady rate, forming the basis for the molecular clock. In the neutral theory of molecular evolution, neutral mutations provide genetic drift as the basis for most variation at the molecular level.

- A nearly neutral mutation is a mutation that may be slightly deleterious or advantageous, although most nearly neutral mutations are slightly deleterious.

Distribution of Fitness Effects

Attempts have been made to infer the distribution of fitness effects (DFE) using mutagenesis experiments and theoretical models applied to molecular sequence data. DFE, as used to determine the relative abundance of different types of mutations (i.e., strongly deleterious, nearly neutral or advantageous), is relevant to many evolutionary questions, such as the maintenance of genetic variation, the rate of genomic decay, the maintenance of outcrossing reproduction as opposed to inbreeding and the evolution of sex and genetic recombination. In summary, the DFE plays an important role in predicting evolutionary dynamics. A variety of approaches have been used to study the DFE, including theoretical, experimental and analytical methods.

- Mutagenesis experiment: The direct method to investigate the DFE is to induce mutations and then measure the mutational fitness effects, which has already been done in viruses, bacteria, yeast, and *Drosophila*. For example, most studies of the DFE in viruses used site-directed mutagenesis to create point mutations and measure relative fitness of each mutant. In *Escherichia coli*, one study used transposon mutagenesis to directly measure the fitness of a random insertion of a derivative of Tn10. In yeast, a combined mutagenesis and deep sequencing approach has been developed to generate high-quality systematic mutant libraries and measure fitness in high throughput. However, given that many mutations have effects too small to be detected. and that mutagenesis experiments can detect only mutations of moderately large effect; DNA sequence data analysis can provide valuable information about these mutations.

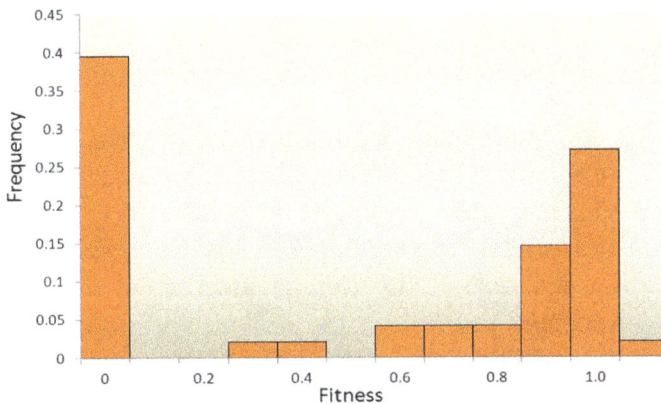

The distribution of fitness effects (DFE) of mutations in vesicular stomatitis virus. In this experiment, random mutations were introduced into the virus by site-directed mutagenesis, and the fitness of each mutant was compared with the ancestral type. A fitness of zero, less than one, one, more than one, respectively, indicates that mutations are lethal, deleterious, neutral, and advantageous.

- Molecular sequence analysis: With rapid development of DNA sequencing technology, an enormous amount of DNA sequence data is available and even more is forthcoming in the future. Various methods have been developed to infer the DFE from DNA sequence data. By examining DNA sequence differences within and between species, we are able to infer various characteristics of the DFE for neutral, deleterious and advantageous mutations. To be specific, the DNA sequence analysis approach allows us to estimate the effects of mutations with very small effects, which are hardly detectable through mutagenesis experiments.

One of the earliest theoretical studies of the distribution of fitness effects was done by Motoo Kimura, an influential theoretical population geneticist. His neutral theory of molecular evolution proposes that most novel mutations will be highly deleterious, with a small fraction being neutral. Hiroshi Akashi more recently proposed a bimodal model for the DFE, with modes centered around highly deleterious and neutral mutations. Both theories agree that the vast majority of novel mutations are neutral or deleterious and that advantageous mutations are rare, which has been supported by experimental results. One example is a study done on the DFE of random mutations in vesicular stomatitis virus. Out of all mutations, 39.6% were lethal, 31.2% were non-lethal deleterious, and 27.1% were neutral. Another example comes from a high throughput mutagenesis experiment with yeast. In this experiment it was shown that the overall DFE is bimodal, with a cluster of neutral mutations, and a broad distribution of deleterious mutations.

Though relatively few mutations are advantageous, those that are play an important role in evolutionary changes. Like neutral mutations, weakly selected advantageous mutations can be lost due to random genetic drift, but strongly selected advantageous mutations are more likely to be fixed. Knowing the DFE of advantageous mutations may lead to increased ability to predict the evolutionary dynamics. Theoretical work on the DFE for advantageous mutations has been done by John H. Gillespie and H. Allen Orr. They proposed that the distribution for advantageous mutations should be exponential under a wide range of conditions, which, in general, has been supported by experimental studies, at least for strongly selected advantageous mutations.

In general, it is accepted that the majority of mutations are neutral or deleterious, with advantageous mutations being rare; however, the proportion of types of mutations varies between species. This indicates two important points: first, the proportion of effectively neutral mutations is likely to vary between species, resulting from dependence on effective population size; second, the average effect of deleterious mutations varies dramatically between species. In addition, the DFE also differs between coding regions and noncoding regions, with the DFE of noncoding DNA containing more weakly selected mutations.

By Impact on Protein Sequence

- A frameshift mutation is a mutation caused by insertion or deletion of a number of nucleotides that is not evenly divisible by three from a DNA sequence. Due to the triplet nature of gene expression by codons, the insertion or deletion can disrupt the reading frame, or the grouping of the codons, resulting in a completely different translation from the original. The earlier in the sequence the deletion or insertion occurs, the more altered the protein produced is. (For example, the code CCU GAC UAC CUA codes for the amino acids proline, aspartic acid, tyrosine, and leucine. If the U in CCU was deleted, the resulting sequence would be CCG ACU ACC UAx, which would instead code for proline, threonine, threonine, and part of another amino acid or perhaps a stop codon (where the x stands for the following nucleotide).) By contrast, any insertion or deletion that is evenly divisible by three is termed an *in-frame mutation*.

- A point substitution mutation results in a change in a single nucleotide and can be either synonymous or nonsynonymous.

  ○ A synonymous substitution replaces a codon with another codon that codes for the same amino acid, so that the produced amino acid sequence is not modified. Synonymous mutations occur due to the degenerate nature of the genetic code. If this mutation does not result in any phenotypic effects, then it is called silent, but not all synonymous substitutions are silent. (There can also be silent mutations in nucleotides outside of the coding regions, such as the introns, because the exact nucleotide sequence is not as crucial as it is in the coding regions, but these are not considered synonymous substitutions).

  ○ A nonsynonymous substitution replaces a codon with another codon that codes for a different amino acid, so that the produced amino acid sequence is modified. Nonsynonymous substitutions can in turn be classified as nonsense or missense mutations:

    ◊ A missense mutation or changes a nucleotide is to cause substitution of a different amino acid. This in turn can render the resulting protein nonfunctional. Such mutations are responsible for diseases such as Epidermolysis bullosa, sickle-cell disease, and SOD1-mediated ALS. On the other hand, if a missense mutation occurs in an amino acid codon that results in the use of a different, but chemically similar, amino acid, then sometimes little or no change is rendered in the protein. For example, a change from AAA to AGA will encode arginine, a chemically similar molecule to the intended lysine. In this latter case the mutation will have little or no effect on phenotype and therefore be neutral.

    ◊ A nonsense mutation is a point mutation in a sequence of DNA that results in a premature stop codon, or a *nonsense codon* in the transcribed mRNA, and possibly a truncated, and often nonfunctional protein product. This sort of mutation has been linked to different mutations, such as congenital adrenal hyperplasia.

By Inheritance

In multicellular organisms with dedicated reproductive cells, mutations can be subdivided into germline mutations, which can be passed on to descendants through their reproductive cells, and somatic mutations (also called acquired mutations), which involve cells outside the dedicated reproductive group and which are not usually transmitted to descendants.

A mutation has caused this moss rose plant to produce flowers of different colors.
This is a somatic mutation that may also be passed on in the germline

A germline mutation gives rise to a *constitutional mutation* in the offspring, that is, a mutation that is present in every cell. A constitutional mutation can also occur very soon after fertilization, or continue from a previous constitutional mutation in a parent.

The distinction between germline and somatic mutations is important in animals that have a dedicated germline to produce reproductive cells. However, it is of little value in understanding the effects of mutations in plants, which lack dedicated germline. The distinction is also blurred in those animals that reproduce asexually through mechanisms such as budding, because the cells that give rise to the daughter organisms also give rise to that organism's germline. A new germline mutation not inherited from either parent is called a *de novo* mutation.

Diploid organisms (e.g., humans) contain two copies of each gene—a paternal and a maternal allele. Based on the occurrence of mutation on each chromosome, we may classify mutations into three types:

- A heterozygous mutation is a mutation of only one allele.

- A homozygous mutation is an identical mutation of both the paternal and maternal alleles.

- Compound heterozygous mutations or a genetic compound consists of two different mutations in the paternal and maternal alleles.

A wild type or homozygous non-mutated organism is one in which neither allele is mutated.

Special Classes

- Conditional mutation is a mutation that has wild-type (or less severe) phenotype under certain "permissive" environmental conditions and a mutant phenotype under certain "restrictive" conditions. For example, a temperature-sensitive mutation can cause cell death at high temperature (restrictive condition), but might have no deleterious consequences at a lower temperature (permissive condition). These mutations are non-autonomous, as their manifestation depends upon presence of certain conditions, as opposed to other mutations which appear autonomously. The permissive conditions may be temperature, certain chemicals, light or mutations in other parts of the genome. *In vivo* mechanisms like transcriptional switches can create conditional mutations. For instance, association of Steroid Binding Domain can create a transcriptional switch that can change the expression of a gene based on the presence of a steroid ligand. Conditional mutations have applications in research as they allow control over gene expression. This is especially useful studying diseases in adults by allowing expression after a certain period of growth, thus eliminating the deleterious effect of gene expression seen during stages of development in model organisms. DNA Recombinase systems like Cre-Lox Recombination used in association with promoters that are activated under certain conditions can generate conditional mutations. Dual Recombinase technology can be used to induce multiple conditional mutations to study the diseases which manifest as a result of simultaneous mutations in multiple genes. Certain inteins have been identified which splice only at certain permissive temperatures, leading to improper protein synthesis and thus, loss of function mutations at other temperatures. Conditional mutations may also be used in genetic studies associated with ageing, as the expression can be changed after a certain time period in the organism's lifespan.

- Replication timing quantitative trait loci affects DNA replication.

Nomenclature

In order to categorize a mutation as such, the "normal" sequence must be obtained from the DNA of a "normal" or "healthy" organism (as opposed to a "mutant" or "sick" one), it should be identified and reported; ideally, it should be made publicly available for a straightforward nucleotide-by-nucleotide comparison, and agreed upon by the scientific community or by a group of expert geneticists and biologists, who have the responsibility of establishing the *standard* or so-called "consensus" sequence. This step requires a tremendous scientific effort. Once the consensus sequence is known, the mutations in a genome can be pinpointed, described, and classified. The committee of the Human Genome Variation Society (HGVS) has developed the standard human sequence variant nomenclature, which should be used by researchers and DNA diagnostic centers to generate unambiguous mutation descriptions. In principle, this nomenclature can also be used to describe mutations in other organisms. The nomenclature specifies the type of mutation and base or amino acid changes:

- Nucleotide substitution (e.g., 76A>T) — The number is the position of the nucleotide from the 5' end; the first letter represents the wild-type nucleotide, and the second letter represents the nucleotide that replaced the wild type. In the given example, the adenine at the 76th position was replaced by a thymine.

- If it becomes necessary to differentiate between mutations in genomic DNA, mitochondrial DNA, and RNA, a simple convention is used. For example, if the 100th base of a nucleotide sequence mutated from G to C, then it would be written as g.100G>C if the mutation occurred in genomic DNA, m.100G>C if the mutation occurred in mitochondrial DNA, or r.100g>c if the mutation occurred in RNA. Note that, for mutations in RNA, the nucleotide code is written in lower case.

- Amino acid substitution (e.g., D111E) — The first letter is the one letter code of the wild-type amino acid, the number is the position of the amino acid from the N-terminus, and the second letter is the one letter code of the amino acid present in the mutation. Nonsense mutations are represented with an X for the second amino acid (e.g. D111X).

- Amino acid deletion (e.g., ΔF508) — The Greek letter Δ (delta) indicates a deletion. The letter refers to the amino acid present in the wild type and the number is the position from the N terminus of the amino acid were it to be present as in the wild type.

Prion Mutations

Prions are proteins and do not contain genetic material. However, prion replication has been shown to be subject to mutation and natural selection just like other forms of replication. The human gene PRNP codes for the major prion protein, PrP, and is subject to mutations that can give rise to disease-causing prions.

Somatic Mutations

A change in the genetic structure that is not inherited from a parent, and also not passed to offspring, is called a somatic mutation. *Somatic mutations are not inherited because they do not affect the germline. These types of mutations are usually prompted by environmental causes, such as ultraviolet radiation or any exposure to certain harmful chemicals, and can cause diseases including cancer.*

With plants, some somatic mutations can be propagated without the need for seed production, for example, by grafting and stem cuttings. These type of mutation have led to new types of fruits, such as the "Delicious" apple and the "Washington" navel orange.

Human and mouse somatic cells have a mutation rate more than ten times higher than the germline mutation rate for both species; mice have a higher rate of both somatic and germline mutations per cell division than humans. The disparity in mutation rate between the germline and somatic tissues likely reflects the greater importance of genome maintenance in the germline than in the soma.

Amorphic Mutations

An amorph, a term utilized by Muller in 1932, is a mutated allele, which has lost the ability of the parent (whether wild type or any other type) allele to encode any functional protein. An amorphic mutation may be caused by the replacement of an amino acid that deactivates an enzyme or by the deletion of part of a gene that produces the enzyme.

Cells with heterozygous mutations (one good copy of gene and one mutated copy) may function normally with the unmutated copy until the good copy has been spontaneously somatically mutated. This kind of mutation happens all the time in living organisms, but it is difficult to measure the rate. Measuring this rate is important in predicting the rate at which people may develop cancer.

Point mutations may arise from spontaneous mutations that occur during DNA replication. The rate of mutation may be increased by mutagens. Mutagens can be physical, such as radiation from UV rays, X-rays or extreme heat, or chemical (molecules that misplace base pairs or disrupt the helical shape of DNA). Mutagens associated with cancers are often studied to learn about cancer and its prevention.

Hypomorphic and Hypermorphic Mutations

A hypomorphic mutation is a replacement of amino acids that would hinder enzyme activity, which would reduce the enzyme level but not to the point of complete loss. Usually, hypomorphic mutations are recessive, but haploinsufficiency causes some alleles to be dominant.

A hypermorphic mutation changes the regulation of the gene and causes it to over-produce the gene produce causing a greater than normal enzyme levels. These type of alleles are dominant gain of function type of alleles.

Mutationism

Mutationism refers to historical and contemporary views of evolution that emphasize the role of mutation and that are understood as alternatives to Darwinism. In the mutationist view, change may occur in discrete jumps (i.e., gradualism is not assumed), mutation is seen as the source of novelty (while selection is not seen as creative), and the direction of evolution is understood to reflect both mutation and selection.

By 1909, "mutationism" was referenced as a distinctive rival of Darwinism, following the discovery and naming of "mutation" by Hugo De Vries. At that time, Darwin's mechanism of natural selection was understood to rely on hereditary blending of abundant continuous variations. However, Wilhelm Johannsen's "pure line" experiments appeared to refute this mechanism. Using true-breeding varieties of beans, each with a different size of seeds, Johannsen showed selection could be used to sort out different varieties, but selection within pure lines would not produce evolutionary changes, even though pure lines continued to generate the kind of abundant variations that Darwinians saw as the fuel for evolution.

At the same time, Hugo de Vries's careful studies of wild variants of *Oenothera lamarckiana*, initially published in 1901,showed that distinct new forms could arise suddenly in nature, and could be propagated for many generations without dissipation or blending. De Vries used the term "mutation" to refer to these dramatic variants, as well as any smaller variations that arise via a sudden event and are inherited stably (as distinct from unstable quantitative fluctuations). Thereafter, "mutationism" came to refer to a view of evolution emphasizing that inheritance is discrete, and that hereditary variants originate in discrete events of mutation, which have a role in shaping the outcome of evolution. The terms "saltationism" and "Mendelism" appear to be used similarly. This was contrasted with a Darwinian view in which variation is continuous, and hereditary variations emerge as masses of infinitesimal effects that are shaped by selection.

These and other early discoveries of genetics are often framed relative to a controversy between, on the one hand, early geneticists— the "Mendelians"— including William Bateson, Wilhelm Johannsen, Hugo de Vries, Thomas Hunt Morgan, and Reginald, who advocated Mendelism and mutation, and were understood as opponents of Darwin's original view, and the biometricians and their allies, who opposed Mendelism and were more faithful to Darwin's original vision. The dispute was resolved, and the eclipse of Darwinism ended, with the rise of the "Modern Synthesis" or modern neo-Darwinism.

Neodarwinist View

While historians agree on the above generalities, they disagree on the nature of the critique of Darwinism offered by geneticists, their alternative view of evolution, and the Darwinian restoration in the modern evolutionary synthesis. In the classical view, early geneticists, upon the discovery of mutation, mistakenly assume that all mutations are large, exaggerate their importance, reject Darwin's idea of selection, and imagine an entire theory of evolution based on dramatic mutations, in which new species were created in a single step; this view gains popularity and temporarily eclipses Darwin's theory; it takes decades for scientists to come to their senses and see that genetics and selection may be combined in a way that allows gradual evolution consistent with Darwin's original vision. The classical view, which emerged in Modern Synthesis writings, presents the Modern Synthesis as the received view and mutationism as an obvious error; the decades-long delay in synthesizing genetics and Darwinism is seen as an "inexplicable embarrassment".

The nature of this synthesis was that, due to recombination in the diverse gene pool of a population, selection can shift the phenotype well beyond its initial limits, essentially creating a new form without any mutations. This argument for "the effectiveness of selection" is the centerpiece of Province's seminal history of theoretical population genetics. Thus, in the classical view, scientists who look to distinctive individual mutations for the origin of novelty, rather than to selection smoothly shifting an entire distribution (of many infinitesimal variations) to fit external conditions, are seen as advocates of a failed theory. Mutationism (saltationism), along with Lamarckism and orthogenesis, are presented as anti-Darwinian "blind alleys" separate from the main line leading from Darwin to the present.

Historical View

Though the classical view is commonly repeated by scientists, it is not clear how strongly historians would defend it against revisionist claims. Gayon's Darwinism's Struggle for Survival argues that Darwin's conception of natural selection was incompatible with Mendelian genetics, and that early geneticists are responsible for "the most important event in the history of Darwinism: the Mendelian reconstruction of the principle of selection" as a force that shifts frequencies of discretely inherited types. That is, Gayon places the early geneticists, not in a blind alley, but in the middle of the genetic reformation of evolutionary thinking, validating their main complaints against Darwin's non-Mendelian view of a creative force of selection.

Provine's seminal history of theoretical population genetics presents a version of the classical narrative, yet in the 2001 reprinting of the same work, Provine rejects the inevitability of the Modern Synthesis (which "came unraveled"), and questions both the sufficiency and the logical coherence of its selection-and-recombination-driven "gene pool" view of population genetics. Provine lists a variety of post-Synthesis findings that are not consistent with the synthetic view of population genetics presented in his 1971 analysis.

The most sympathetic view of the mutationists is given by Stoltzfus and Cable, who argue that early geneticists accepted selection, allowed gradual evolution, and laid the conceptual foundations for the genetical view of evolution that prevails today. They argue that the fusion of Darwinism and genetics that Provine describes achieved premature acceptance, but later proved untenable and is no longer the foundation of contemporary evolutionary thinking, which is better aligned with the "Mendelian-mutationism" of early geneticists.

In summary, classical historiography holds that mutationism is a failed view in which the role of mutation in evolution is misperceived, while the role of selection is wrongly dismissed. In revisionist views, the mutationists accept both mutation and selection, and assign them the same qualitative roles they are assigned today, though perhaps not the same quantitative emphasis. At the time of the Darwin centennial in Cambridge in 1909, Mutationism and Lamarckism were contrasted with Darwin's "Natural Selection" as competing ideas; 50 years later, at the University of Chicago centennial of the publication of *On the Origin of Species*, mutationism (like Lamarckism) was no longer seriously considered. Nevertheless, after another 50 years, evolutionary biologists are reconsidering the mutationist view.

Neutral Mutations

Neutral mutation is a mutation that has no selective advantage or disadvantage. Considerable controversy surrounds the question of whether such mutations can exist.

When the idea of a constant molecular clock first emerged, it was thought that the predominant evolutionary force underlying amino acid or nucleotide substitutions was natural selection. Following this line of thinking, a constant molecular clock would indicate that adaptive substitutions in different species occur constantly over time. However, it is hard to explain how adaptive substitutions would occur in such a clock-like manner. Theoretically, the fates of adaptive mutations are determined by several evolutionary parameters, such as the strength of the selective advantage of that mutation, the size of the effective population, and adaptive mutation rates. These parameters are likely to differ between species, and even within a species, depending upon specific mutations and their interactions with environments.

Instead, Kimura proposed that most changes at the molecular level have little functional consequences, or are 'neutral'. If a mutation has no fitness consequence, its fate in the population is determined completely by random chance. This means that we cannot predict whether a specific neutral mutation will eventually be fixed in the population. However, the *rate* at which neutral substitutions occur in the population can be predicted because it depends upon a single parameter, namely the mutation rate.

Let's imagine a population with N number of haploid individuals. If neutral mutations occur at rate u per individual per generation, the total number of mutations in one generation will be N times u. Since all these new mutations are neutral, their fates are completely determined by chance. In other words, all mutations have equal chance of reaching fixation (which leads to a 'substitution'). The probability that each new neutral mutation will reach fixation, given that a substitution occurred, is simply $1/N$. The rate of substitutions is calculated as the number of new mutations in each generation (Nu) multiplied by the probability each new mutation reaches fixation ($1/N$), which equals u. In other words, for neutral mutations, the rate of substitution is equal to the rate of mutation.

Almost all the controversies at the heart of debates over neutral molecular clocks stem from the question of what the major sources of mutations are. This question is directly relevant to understanding patterns of mutation, which are the ultimate source of evolutionary change and genetic disease. Furthermore, understanding how mutation rates vary between lineages and within genomes is a fundamental question in comparative genomics, which aims to use sequence comparisons to identify genomic regions that are functionally important.

So what determines neutral mutation rates? One of the most important contributors to neutral molecular clocks is lineage-specific variation in generation times. From early on, the idea of a constant neutral molecular clock was perceived as being at odds with the molecular mechanisms of germline mutation. It has long been considered that most mutations arise from errors in DNA replication in gremlines. Since mutations occur when germline DNA is replicated for the next generation, they should accumulate in proportion to the number of generations, rather than the absolute amount of time. Therefore, if we compared the numbers of substitutions that have accumulated in two lineages since their divergence, the lineage with longer generation time, having undergone fewer DNA replication events, would harbor fewer substitutions compared to the lineage with the shorter generation time. Consequently, the molecular clock should run more slowly in species with longer generation times. This idea is referred to as the 'generation-time effect'.

Point Mutation

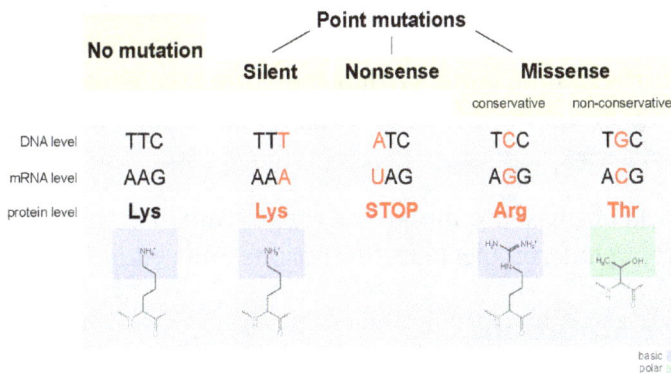

	No mutation	Silent	Nonsense	Missense conservative	non-conservative
DNA level	TTC	TTT	ATC	TCC	TGC
mRNA level	AAG	AAA	UAG	AGG	ACG
protein level	Lys	Lys	STOP	Arg	Thr

Point mutation is a mutation in which change within a gene in which one base pair in the DNA sequence is altered. Point mutations are frequently the result of mistakes made during DNA replication, although modification of DNA, such as through exposure to X-rays or to ultraviolet radiation, also can induce point mutations.

There are two types of point mutations: transition mutations and transversion mutations. Transition mutations occur when a pyrimidine base (i.e., thymine [T] or cytosine [C]) substitutes for another pyrimidine base or when a purine base (i.e., adenine [A] or guanine [G]) substitutes for another purine base. In double-stranded DNA each of the bases pairs with a specific partner on the corresponding strand—A pairs with T and C pairs with G. Thus, an example of a transition mutation is a GC base pair that replaces a wild type (or naturally occurring) AT base pair. In contrast, transversion mutations occur when a purine base substitutes for a pyrimidine base, or vice versa; for example, when a TA or CG pair replaces the wild type AT pair.

At the level of translation, when RNA copied from DNA is converted into a string of acids during protein synthesis, point mutations often manifest as functional changes in the final protein product. Thus, there exist functional groupings for point mutations. These groupings are divided into silent mutations, missense mutations, and nonsense mutations.

Missense Mutation

A missense mutation is when the change of a single base pair causes the substitution of a different amino acid in the resulting protein. This amino acid substitution may have no effect, or it may render the protein nonfunctional.

The necessity to understand the effects of missense mutations becomes clear when one considers genetic diseases. It is known that the substitution of only one residue in a protein sequence can be related to a number of pathological conditions such as Alzheimer's, Parkinson's and Creutzfeldt-Jakob's diseases.[6] It is also well known that accumulation of autosomal mutations can lead to cancers, and that hereditary diseases are caused by one or more germline mutations. The analysis of the impacts of missense mutations advances our understanding of the relationships between protein structure and function, and allows us to decipher the mechanisms of the effects of disease mutations and thereby pathogenesis.With personalized medicine on the not-so-distant horizon, it is swiftly becoming essential to understand the process by which genetic mutations lead to disease.

The plausible effects of missense mutations range from affecting the macromolecular stability to perturbing macromolecular interactions and cellular localization.

Point nonsense Mutation

A nonsense mutation is the substitution of a single base pair that leads to the appearance of a stop codon where previously there was a codon specifying an amino acid. The presence of this premature stop codon results in the production of a shortened, and likely nonfunctional, protein.

Frameshift Mutation

A frameshift mutation is a genetic mutation caused by a deletion or insertion in a DNA sequence that shifts the way the sequence is read. A DNA sequence is a chain of many smaller molecules called nucleotides. DNA (or RNA) nucleotide sequences are read three nucleotides at a time in units called codons, and each codon corresponds to a specific amino acid or stop signal. During translation, the sequence of codons is read in order from the nucleotide sequence to synthesize a chain of amino acids and form a protein. Frameshift mutations arise when the normal sequence of codons is disrupted by the insertion or deletion of one or more nucleotides, provided that the number of nucleotides added or removed is not a multiple of three. For instance, if just one nucleotide is deleted from the sequence, then all of the codons including and after the mutation will have a disrupted reading frame. This can result in the incorporation of many incorrect amino acids into the protein. In contrast, if three nucleotides are inserted or deleted, there will be no shift in the codon reading frame; however, there will be either one extra or one missing amino acid in the final protein. Therefore, frameshift mutations result in abnormal protein products with an incorrect amino acid sequence that can be either longer or shorter than the normal protein.

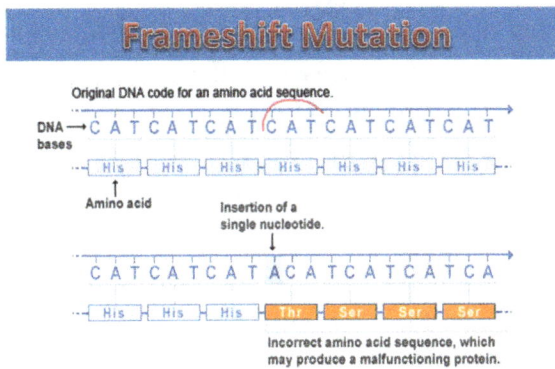

Suppressor Mutation

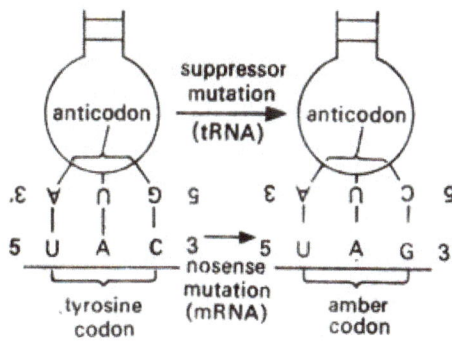

When the effect of one mutation is suppressed by another mutation, the latter is called a suppressor mutation. For instance, AABB may be wild, and aaBB a mutant; then if aabb, is also a wild type, the mutation in 'B' will be called suppression mutation, because the effect of 'a' is suppressed by 'b' Mechanism of this suppression effect was not known for some time, but now its mechanism is largely known. These effects can be conveniently classified into:

1. Intragenic suppression,

2. Intergenic suppression.

Intragenic Suppression

Suppression of a mutation at one site within a gene can be achieved by mutation at another site within the same gene. For instance, a mutant A46 that produces inactive tryptophan syntheses enzyme, due to substitution of glutamic acid for glycine, at position 210 may be corrected or suppressed due to mutation A446 involving substitution of cysteine in place of tyrosine at position 174. These two mutants A46 and A446 independently give rise to inactive enzyme, but if present together give rise to active enzyme, thus suppressing the mutant effect of each other. In many other combinations, double mutants do not show any suppression effect. The suppression effect really depends on whether or not the second alteration, can restore the 3-D configuration of the enzyme. There are a number of known examples of such intragenic or intracistronic suppression or complementation. The reader is advised to consult an advanced book for further details in this connection.

Intergenic suppression, which means suppression of mutation in one gene due to mutation in another gene, can be achieved mainly due to alteration in transfer RNA (tRNA). The suppression may also be effective by alteration brought about in ribosome conformation.

There are atleast four types of such suppression mutations, which are discussed below-

Nonsense suppression and mutant tRNA. Sometimes due to a change in anticodon of a specific tRNA, it may develop the capacity to read the mutant terminating codon (amber or UAG, ochre or UAA and opal or UGA mutants). Since these nonsense mutations cause premature termination of polypeptide chain, they can be suppressed if a mutant tRNA can read the terminating codon and substitute an amino acid at the location of termination codon. Mutant tRNAs are known which are capable of recognizing termination codons and insert amino acids like tyrosine, serine or glutamine to restore the function of protein. However these mutant tRNAs may also read the normal stop signals of proteins thus causing elongation of protein beyond its normal termination thus causing disturbances. To prevent this disturbance, double stop signals are often available in mRNA, so that if mutant tRNA reads one, the other may still bring about termination. Moreover, two genes may code for same tRNA, so that if one (minor gene) mutates, and gives rise to suppressor tRNA the other (major gene) still functions and gives rise to normal tRNA.

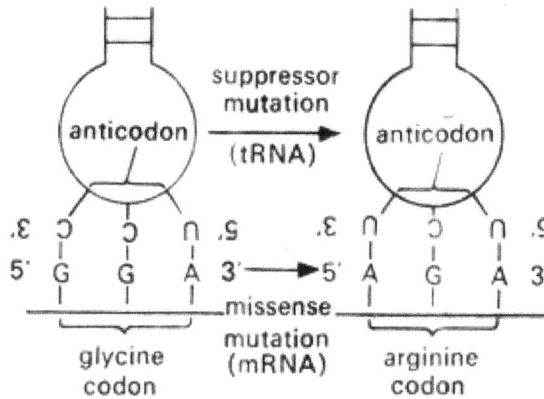

Figure: Suppressor mutation involving a change in one base of anticodon of tyrosine tRNA leading to reading of termination amber codon UAG, which resulted from nonsense mutation in tyrosine codon UAC thus leading to suppression.

Transfer RNA (tRNA) mediated UAG suppression was first observed, when in coat protein of phage R 17, amber codon 5'UAG3' could be read by a tyrosine suppresser tRNA, in which the anticodon had mutated from 5'GUA3' to 5'CUA3'. There are also other examples of this type of suppression.

Missense Suppression and Mutant tRNA

Missense Mutations

Will lead to synthesis of a protein with a wrong amino acid at a particular site. Such a mutation can also be suppressed like nonsense mutations, by mutant tRNA, which will misread the codon and thus substitute a correct amino acid, although the mutant codon codes for a different amino acid. For instance, in mutant A36 in tryptophan synthetase of E. coli, glycine (coded by GGA) is replaced by arginine (coded by AGA). A suppresser mutant may have tRNA, depositing glycine for codon AGA actually meant for arginine.

Frameshift Mutations and Mutant tRNA

It has been shown that a frameshift mutation due to a deletion can be suppressed by mutant tRNA having an anticodon of two bases rather than three (doublet anticodon rather than triplet). Similarly, addition frameshift mutations can be suppressed by mutant tRNAs each having a quadruplet anticodon instead of a triplet. Mutant tRNAs with doublet and quadruplet anticodons have actually been detected in biological systems.

Suppression and Ribosomal Mutations

Mutations in some ribosomal proteins may cause a change in ribosome structure, so that it may now misread the mRNA codons. One such mutation, called *ram* (ribosomal ambiguity) is capable of reading all the three termination codons, thus causing suppression of nonsense mutations. Some frameshift mutations may also be suppressed by these *ram* mutants.

Dynamic Mutation

The term 'dynamic mutation' was introduced to distinguish the unique properties of expanding, unstable DNA repeat sequences from other forms of mutation. The past decade has seen dynamic mutations uncovered as the molecular basis for a growing number of human genetic diseases and for all of the characterized 'rare' chromosomal fragile sites. The common properties of the repeats in different diseases and fragile sites have given insight into this unique form of DNA instability. While the dynamic mutation mechanism explains some unusual genetic characteristics, unexpected findings have raised new questions and challenged some assumptions about the pathways that lead from mutation to disease.

The process of dynamic mutation is affected by a variety of elements and factors that have been divided into those directly associated with the expanding repeat (cis-acting elements) and those (trans-acting factors) whose interaction with the repeat contributes to its instability. Examples of cis-acting factors include the copy number and the composition of the repeat (whether it is perfect or interrupted). In general, higher copy number alleles that are free of interruptions (perfect repeats) are more unstable than those of lower copy number and contain interruptions (imperfect repeats). While there is indirect evidence for the existence of trans-acting factors (such as parental gender bias in repeat instability), the identity of these factors remains elusive. Candidates include the proteins involved in DNA replication such as FEN1 (Rad27), which is known to have a role in Okazaki fragment metabolism.

Dynamic Mutation Undergoing Repeat Sequences

The first expanded repeats to be identified were the trinucleotides CCG/CGG and CAG/CTG. Initially this was taken as evidence that only trinucleotide repeats could undergo this form of mutation (which was sometimes referred to as trinucleotide repeat expansion). Furthermore, since these two repeats could form secondary structures, it was assumed that this was also a necessary condition for repeat instability. Subsequently, six of the ten possible trinucleotide repeats have been found to exhibit repeat expansion in humans, and, more recently, 5 and 6 bp microsatellite repeats and 12, 33 and 42 bp minisatellite repeats have also been observed to undergo repeat copy number expansion.

Founder effects are observed in association with dynamic mutation as a higher frequency of particular alleles in the affected population, compared with the unaffected population, suggesting that these alleles are 'at-risk' for mutation. The likelihood is that the initial mutation on these alleles is a rare event that increases the instability of the repeat, setting in process the subsequent expansion of these few founder muta-tions. Founder effects are a common feature of dynamic mutation and in many cases common flanking marker haplotypes provide evidence of a relationship between the expanded alleles and the longest normal (usually uninterrupted) alleles in the popula-tion. Therefore a common molecular mechanism appears to be responsible.

At the *FRAXA* locus, detailed haplotyping studies, together with sequence analysis of different repeat alleles, have provided evidence in support of the view that certain al-leles are predisposed to instability. Similar analyses in other populations suggest that such differences are likely to be population-specific rather than indicative of a molecu-lar pathway predisposed by *cis*-acting elements in and around the *FRAXA* repeat. One possible explanation for the inconsistency here is that chance has a role to play and that the differences between populations merely reflect the different rare initial events that have precipitated the dynamic mutation process. Given that there are multiple possi-bilities for generating an unstable repeat then different populations are likely to have different combinations and/or frequencies of these unstable alleles. Thus, in terms of understanding the mutation process, caution is needed when extrapolating the results observed in one population to the human species as a whole.

Factors Contributing to Repetition in DNA Expansion

Gender bias in the instability of repeats during their transmission from parent to off-spring is a common property of dynamic mutations. Most cases of the congenital form of myotonic dystrophy are maternally inherited, while the juvenile onset Huntington disease is generally from paternal inheritance. The gender of the germ line clearly con-tributes to these biases; however, the identity of the factors involved remains obscure.

Other factors that interact in some way with the repeat are also thought to contribute to repeat instability. A variety of circumstantial evidence (particularly from studies in model systems) suggests that Okazaki fragments may have a role to play in the expansion of DNA repeats. Several of the repeats (e.g. CCG in *FRAXA* and CAG in *DM*) exhibit a bimodal distribution of instability with the boundary between small changes and large changes in copy number approximating Okazaki fragment length. In *Escherichia coli* (which has a substantially longer Okazaki fragment length), this boundary is increased in copy number again to that approximating Okazaki fragment length. The secondary structure of CAG/CTG and CGG repeat sequences has been found to inhibit Okazaki flap endonuclease (FEN1) binding and cleavage. In addition, the FEN1 homologue, Rad27, has a role to play in DNA repeat instability in yeast. Yeast Rad27 mutants are prone to both DNA breakage and copy number instability of repeat sequences.

Contradicting the proposed role of DNA replication, Kennedy and Shelbourne found dramatic expansion of the CAG repeat in the striatum of a transgenic mouse HD mod-el. These neurons are presumably no longer undergoing cell division, indicating that expansion is occurring independently of DNA replication. Similarly, Kotvun and Mc-Murray have found that germ-line trinucleotide repeat sequence expansion in transgenic mice also proceeded in the absence of DNA replication via gap repair. Perhaps, given that DNA instability is occurring at the repeat sequence in the absence of replication, some form of transcriptional healing may be contributing to the repeat expansion process. Transcriptional healing has been invoked in the fragile site expression of tandemly repeated small RNA transcripts. In further support of a role for breakage and repair Manley have found that *Msh2* is required for CAG repeat somatic cell instability in an HD transgenic mouse.

Age-dependent somatic cell instability of the CTG repeat sequence has been found in myotonic dystrophy, although here it is not confined to those cells that are affected in the disease and in fact gives a growth advantage to lymphoblastoid cell lines that have undergone expansion.

Pathogenic Pathways

Loss of Function/Haploinsufficiency

Extinction of, or interference with transcription is one means by which expanded repeats cause loss of gene function. Examples are: fragile X syndrome, in which repeat expansion brings about methylation of the *FMR1* promoter region; myoclonus epilepsy (*EPM1*), in which repeat expansion physically separates transcription factors in the cystatin B promoter; and Friedreich ataxia, in which the intronic expanded repeat sequence adopts a 'sticky' DNA structure that interferes with transcription of the frataxin gene. While these resultant diseases exhibit recessive (or X-linked) transmission, the potential exists for dominant transmission where reduction in the level or activity of a rate limiting gene product can appear as haploinsufficiency. This appears to be the case for the *SIX5* gene in myotonic dystrophy, as the expanded CTG repeat is located in the *SIX5* promoter. The reduction in effective 'dosage' of the *SIX5* gene product brought about by extinction of one allele is thought to contribute to some aspects of pathology.

Gain of Function: The Toxic Polyglutamine Hypothesis—Old Concerns about a New Dogma

Numerous expanded repeat sequences are located within coding regions where they translate into expanded polyglutamine tracts in disease alleles. This, together with a substantial body of data demonstrating the toxicity of polyglutamine in cells, has led to the view that the expanded polyglutamine is a common, crucial component in the pathogenesis of these diseases. In support of this view, an antibody, 1C2, raised against the normal length polyglutamine tract of the TATA-box binding protein (TBP), was found to specifically recognize the expanded polyglutamine in several of the expanded repeat diseases, e.g. huntingtin, ataxin 1 and others. This was taken as evidence that the expanded polyglutamine adopted a different conformation that in some way contributed to its toxicity. The context in which the repeat sequence was located was proposed to account for why such an otherwise toxic conformation did not cause problems when located within TBP. This theory is now somewhat flawed, as the expanded polyglutamine tract in some alleles ($n > 50$) of TBP has recently been associated with neurodegenerative disease. It is possible that yet another conformation of the expanded polyglutamine tract is adopted once the repeat sequence gets beyond an even greater threshold.

Nuclear inclusions were found in the affected neurons of HD and SCA1 transgenic mouse models and HD patients, prompting the assertion that these expanded polyglutamine-containing nuclear inclusions cause neuronal dysfunction. Klemment deleted the SCA1 self-association region in transgenic mice and found that these mice still exhibited ataxia and Purkinje cell pathology; however, no evidence of nuclear aggregates was found. Other studies on the brains of individuals affected with Huntington's disease revealed that nuclear aggregates were more commonly found in unaffected neurons than in the vulnerable striatal spiny neurons. Consequently, it has now been proposed that rather than being pathogenic, nuclear inclusions may even be protective against the toxic effects of the expanded polyglutamine.

Further conflicting evidence to a common role for polyglutamine in pathogenesis is that disease alleles of the CAG repeat in spinocerebellar ataxia 6 are well within the normal range for other 'polyglutamine diseases'. Functional assays on expanded CAG alleles of the SCA6 associated gene, *CACNA1A*, demonstrate increased Ca^{2+} transport and thus make it likely that this gain of function is a specific and unique cause of pathogenesis for this disorder. Taken together, these exceptions suggest that the notion of a common single pathogenic pathway involving polyglutamine is unlikely.

As if contradictions with polyglutamine encoding genes are not enough, a growing list of repeat-associated ataxias do not even encode polyglutamine. In *SCA12*, the expanded CAG repeat is located within the 5′-untranslated region (5′-UTR) of the PPP2R2B gene, while in *SCA8* an expanded CUG is located within the 3′-UTR of a transcript (although there is controversy over whether this expansion is causative of neurodegenerative disease). In addition, an intronic expanded 5 bp repeat was recently found to be associated with *SCA10*.

So the question must be asked whether expanded polyglutamine is a necessary and sufficient condition for those diseases in which it has been identified as the disease-causing agent. If it is, then how is the cellular specificity of this toxicity accounted for when at least some of these proteins are widely expressed? One possible explanation for this conundrum comes from the findings of Kennedy and Shelbourne, which reveal affected cell-specific amplification of the unstable CAG repeat in a mouse model of Huntington's disease. Individuals affected with Huntington's disease have also been found to exhibit somatic instability, which was most pronounced in the affected areas of the brain. This suggests that ongoing somatic mutation may contribute to diseases caused by expanded repeats. This expansion may also account for the relationship between germ-line repeat copy number and age-at-onset for these diseases. The time taken for the disease to manifest could represent how long it takes for the disease-associated allele to reach a critical higher copy number threshold in the affected cell population.

What then is the evidence that polyglutamine has a role? The principal data are from transgenic animal model studies, particularly those from *SCA1* transgenic mice. Mutation of the ataxin-1 nuclear localization signal severely diminishes toxicity, whereas abolition of sequences necessary for aggregation does not. Crossing of *SCA1* mice with a transgenic mouse deficient in ubiquitin ligase results in more severe pathology. For *SCA2* and *SCA6*, nuclear localization does not appear to be necessary, suggesting different pathways than that for SCA1 (and HD).

In an HD transgenic mouse model where transcription of the transgene could be artificially turned on or off, expression of the expanded CAG-containing gene was elegantly demonstrated to be necessary for HD-like pathology. This animal model provides the first indication that HD (and hopefully other expanded CAG repeat diseases) is potentially treatable—given that, in those patients at risk, the disease-causing gene can either be switched off or the toxic gene product can be cleared from the sensitive neurons.

Presumably other factors, such as modification of the polyglutamine tract and the protein context in which the polyglutamine is located, contribute to the cellular specificity of toxicity. Examples of where this apparently is or is not the case have been reported. The relative contribution of various intrinsic and extrinsic factors is therefore likely to differ between the different disease genes (and experimental model systems). Genetic screens for factors that modify polyglutamine-generated pathology in *Drosophila* have produced a large number and variety of candidates, and it will be of great interest to see which (if any) of these has a role in human polyglutamine disease.

Bias of ascertainment was invoked by Penrose to account for (and discredit) the observation of anticipation seen by clinical geneticists in myotonic dystrophy families. While expanded DNA repeats now give a molecular basis (and validation) for the phenomenon of anticipation, the relative ease with which expanded CAG repeats can be screened for as a cause of neurodegenerative disease may well have led to an overemphasis of the relative importance of this form of mutation and perhaps contributed to the view that expanded CAG repeats necessarily have a common pathogenic pathway (via their encoded expanded polyglutamines). While there is a substantial body of evidence in support of a role for polyglutamine in some repeat expansion diseases there is a growing list of exceptions and inconsistencies which suggest (at the least) that this is not necessarily a common pathogenic pathway exhibited by all neurodegenerative diseases in which an expanded repeat (let alone an expanded CAG) is the disease-causing mutation.

Further Evidence for Alternative Pathways—Multiple Gene Effects

In myotonic dystrophy there is good evidence to support not only a role for multiple genes, but also multiple pathways in mediating the pathogenic effects of repeat expansion. Three genes (*DMPK*, *SIX5* and *DMWD*) are all located in the immediate vicinity of the DM expanded CTG repeat. At least three distinct pathogenic pathways are thought to contribute to DM.

1. The expanded repeat affects the splicing of the DMPK gene in which it is located.

2. The RNA transcript containing the expanded repeat is both inappropriately compartmentalized and titrates a crucial protein (muscle blind). Expression of the CTG repeat in another muscle-specific transcript is able to cause at least some of the DM phenotype (myotonia and myopathy).

3. The DM CTG repeat is also located in the promoter region of the *SIX5* gene and its expansion also interferes with the expression of this gene. The *SIX5* gene encodes a transcription factor whose concentration is critical for eye, muscle and testicular development. Heterozygous deletion of *Six5* is sufficient to cause ocular cataracts in transgenic mice, and therefore the *SIX5* haploinsufficiency brought about by the DM CTG repeat expansion is likely to contribute to the DM phenotype.

The *FRAXE* chromosomal fragile site is associated with a mild form of mental retardation. The expanded CCG repeat was first initially located in the 5′-UTR of a gene, referred to as *FMR2*. The expression of *FMR2* is extinguished by repeat expansion and subsequent methylation of the CpG island promoter region, in a manner analogous to the silencing of the *FMR1* gene at the *FRAXA* locus in fragile X syndrome. It was therefore thought that the loss of *FMR2* was responsible for the phenotype associated with *FRAXE*. However, an additional gene, *FMR3*, has now been identified that shares the same methylated CpG island. Expression of *FMR3* is also silenced by *FRAXE* full mutation and therefore this gene may also contribute to the phenotype.

Polar Mutation

Polar mutations in *lacZ*

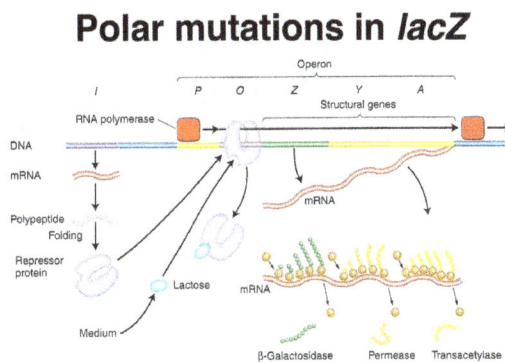

Polar mutation is a mutation that affects the transcription or translation of part of the gene or operon downstream of the mutant site.

The first class of polar mutations involves an "insertion" of a foreign piece of DNA which contains an RNA termination signal. Such a mutation eliminates downstream gene expression since no RNA polymerase will read through the "new" termination signal to reach those genes. One therefore typically sees a complete cessation of downstream mRNA synthesis. Gene expression typically refers to the act of transcription of the gene in question.

The second class of polar mutations are those whose primary effect is the premature termination of protein synthesis. This often leads indirectly to a partial reduction of downstream mRNA synthesis. Such polarity requires the presence of a functional Rho protein, which is normally involved in transcription termination at the end of operons. A model explaining the mechanism by which premature protein stop signals, termed nonsense mutations, Rho factor causes RNA polymerase to terminate transcription downstream from a site of translation termination if the following conditions exist:

i.   A region of the mRNA, lacking ribosomes because of the upstream nonsense signal, contains a site that activates the ATP synthase activity of Rho, probably by binding Rho. There is *in vitro* evidence that such sites are C- rich and G-poor (in the mRNA) and that an approximately 12-base spacing of C residues may be important, though a precise consensus is unclear. This latter number is intriguing since biochemical characterization suggests that Rho complexes with about 13 nucleotides per protein subunit of the hexamer.

ii. At some distance downstream (deliberately vague, but probably not important) from this site, the activated Rho "catches up" with RNA polymerase (at a polymerase pause site?) and causes it to terminate transcription.

This effect requires that there has not been translation initiation, proper or improper, between the "upstream site" and the termination region, or else ribosomes will mask the sites on the mRNA. It seems that termination sites occur fairly frequently and that the "upstream sites" may be the limiting factors in determining where polarity occurs.

The model is the following: if there is untranslated, unstructured mRNA, then there is a probability that Rho will bind there (the probability will be bigger as the "open" mRNA gets longer and when it has sequences that are preferred by Rho). Having bound, there is a probability that it will catch up with the transcribing RNAP and cause termination

Adaptation

A biological adaptation is any structural (morphological or anatomical), physiological, or behavioral characteristics of an organism or group of organisms (such as species) that make it better suited in its environment and consequently improves its chances of survival and reproductive success. Due to individual phenotypic plasticity (variability), individuals will be more or less successful. Some adaptations may improve reproductive success of the population, but not a particular individual, such as seen in altruistic behavior in social insects.

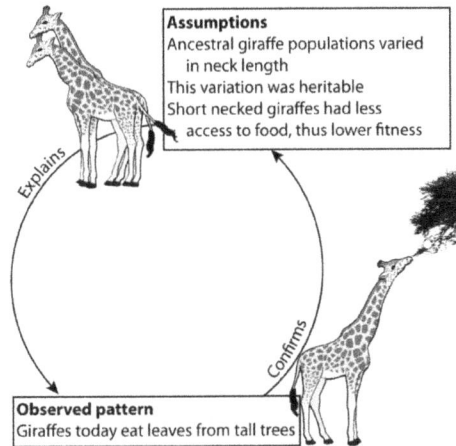

Organisms that are adapted to their environment are able to:

- Secure food, water, and nutrients.

- Obtain air, warmth, and spaces.

- Cope with physical conditions such as temperature, light, and heat.

- Defend themselves from their natural enemies.

- Reproduce and rear offspring.

- Respond to changes around them.

Adaptation occurs in response to changes in the environment, life style, or relationship to other organisms. Environmental dynamicity, voluntary or compelled shifting of habitat and human activities may put organisms in a new niche or in environmental stresses or pressures. In such circumstances, the organisms require characteristics suitable to the new situation. Organisms that are not suitably adapted to their environment will either have to move out of the habitat or die out. The term die out in the context of adaptation means that the death rate over the entire population of the species exceeds the birth rate for a long enough period for the species to disappear.

While adaptations provide for the individual purpose of the organism—survival, reproduction, development, maintenance—these same characteristics provide diversity and add to human fascination with, and enjoyment of, nature. Furthermore, while adaptations often are seen as a static set of suitable characteristics, in reality the process of developing adaptations is a dynamic process. Whether envisioned as the product of design or natural selection, or natural selection on the micro evolutionary level and design for macro evolutionary changes, the reality is that new adaptations are needed when organisms encounter new environments, and such have arisen for millions of years.

In some extreme conditions, it is possible for the previous adaptation to be poorly selected, the advantage it confers over generations decreasing, up to and including the adaptation becoming a hindrance to the species' long–term survival. This is known as maladaptation.

There is a great difference between adaptation and acclimation or acclimatization. The process of developing adaptations occurs over many generations; it is a population phenomenon involving genetics and is generally a slow process. Acclimation or acclimatization, on the other hand, generally occurs within a single lifetime or instantly and deals with issues that are less threatening. For example, if a human being were to move to a higher altitude, respiration and physical exertion will become a problem. However, after spending a period of time under the high altitude conditions, one may acclimatize to the reduced pressure, the person's physiology may function normally, and the change will no longer be noticed.

Types of Adaptation

Adaptations can be structural, physiological, or behavioral. Structural adaptations are special body parts of an organism that help it to survive in its natural habitat (e.g., skin color, shape, body covering). Physiological adaptations are systems present in an organism that allow it to perform certain biochemical reactions (e.g., making venom, secreting slime, being able to keep a constant body temperature). Behavioral adaptations are special ways a particular organism behaves to survive in its natural habitat (e.g., becoming active at night, taking a certain posture).

Based on the habitats for which organisms develop adaptations, adaptations can be categorized into 3 fundamental types, namely aquatic, terrestrial, and volant (flying), each of which can be further divided into many subtypes.

Aquatic Adaptation

Aquatic adaptations are found in those plants and animals that live in water habitats: fresh water, brackish water, and sea water. For example, fresh water organisms develop features to prevent the entry of excess water or processes to drain excess water regularly. On the contrary, marine organisms face scarcity of water due to hypertonic (salt concentration higher than that of body fluid) sea water. So, they have mechanisms to retain water and excrete excess salts that enter in water intake. Aquatic plants may be emergent rooted plants (e.g., reeds), submersed rooted plants (e.g., *Hydrilla*), planktons (e.g., diatoms) or floating plants (e.g., water hyacinth). Similarly, aquatic animals may be benthic, occurring at the bottom of a water body, or pelagic, occurring in the water body itself. The animals may live partially or permanently in water. Thus they may range from non–specialized to very highly specialized water dwellers.

Primarily aquatic animals (e.g., fishes) show not a single terrestrial feature, whereas secondarily aquatic animals (whales, dolphins) possess terrestrial respiration through lungs, and some must visit land for laying eggs (e.g., turtle). Partially water dwelling animals demonstrate amphibious adaptations with double features both for land and water (e.g., frogs, salamanders), or mostly terrestrial features and only some basic aquatic adaptations (e.g., duck).

Some characteristic aquatic adaptations are:

- *Body contour* is spindle shaped and *streamlined*. For this, the head is elongated into rostrum or similar structure, neck is short, external ears (pinnae) are reduced, and tail is laterally or dorso–ventrally compressed.

- Usually marine animals are excessively large (e.g., whale), because of the *buoyancy* of the salt water.

- Organs of locomotion and balancing vary greatly among the aquatic animals; fishes use paired and unpaired *fins*, whales and turtles have their limbs modified into *paddles*, in some others, hands and feet are *webbed*.

- Skin of most aquatic forms is rich in mucous glands to make it slippery. Fishes are equipped with *dermal scales* as well. Aquatic mammals have reduced or absent hair and skin glands (oil and sweat glands). In compensation, they have a fatty layer below the skin known as *bubbler*. Besides insulating the body, it also helps in flotation.

- Primarily aquatic animals are capable of utilizing dissolved oxygen in the water for respiration through general body surface, internal or external *gills*, and so forth. However, secondarily aquatic forms respire atmospheric air through lungs; nostrils are located at the apex of the head.

- In fish, the hollow outgrowth of the alimentary canal, called *air bladder*, functions as an organ of flotation and accessory respiratory organ as it is filled with air. In whales and other mammals, extraordinarily massive lungs and closable nostrils serve this purpose.

- Fishes have *lateral line* systems extending the whole length of the body. It contains neuromast organs, which act as rheoreceptors (pressure receptors).

Terrestrial Adaptation

Terrestrial adaptations are exhibited by the plants and animals living in land habitats. As there are varied types of land habitats, the adaptations shown by organisms also are of diverse kinds.

Fossorial Adaptation

This adaptation occurs in the animals leading a subterranean mode of life. They are equipped with digging organs and they dig for food, protection, or for shelter. Zoologically, they tend to be primitive and defenseless. The adaptational features are:

- The body contour is cylindrical, spindle–shaped, or fusiform (e.g., earthworms, moles, badgers) so as to reduce resistance in subterranean passage.
- The head is small and tapers anteriorly to form a burrowing snout.
- Neck and pinnae are reduced to avoid obstruction in quick movement through the holes. In some, tail is also shortened.
- The eyes remain small and non–functional.
- Limbs are short and strong. Paws are broad and stout with long claws and some extra structures for digging. In *Gryllotalpa* (mole–cricket), the forelegs are modified into digging organs.

Cursorial Adaptation

This is adaptation involving "running" and is required by those organisms living in grassland habitats, since the lack of hiding places means fast running is an important means of protection from the enemies there. Horses, zebras, deer, and so forth show this adaptation, with following modifications:

- The neck is reduced and the body is streamlined, this will reduce the air resistance while running.
- The bones of palms (carpals, metacarpals) and soles (tarsus, metatarsus) become compact and are often fused to form cannon bone.
- The forearm bone ulna and shank bone fibula are reduced.
- Distal segments of both limbs, such as radius, tibia, and canon bones, are elongated to increase the length of the stride.
- Movement of the limbs is restricted to a fore-and-aft plane.

Arboreal Adaptation

This is also known as scansorial adaptation and is found in animals that live in trees or climb on rocks and walls. The features enabling them to be best suited in the habitat are:

- The chest, girdles, ribs, and limbs are strong and stout.
- Feet and hands become *prehensile* (catching) with opposable digits (e.g., primates, marsupials). Sometimes, the digits are grouped as 3 digits and 2 digits in the *syndactyly* (e.g., *Chameleon*). For facilitating the clinging, some have

elongated claws (e.g., squirrels), while others bear rounded adhesive pads at the tip of the digits (e.g., the tree frog *Hyla*). In the wall lizard *(Hemidactylus)*, there are double rows of *lamellae* in the ventral side of digits for creating vacuum to cling. This enables the animals to move even on the smooth vertical surfaces.

- Often the tail becomes prehensile as well (e.g., chameleon, monkeys).

Desert Adaptation

Desert adaptations are for the mode of life in extreme terrestrial habitats. Desert plants *(xerophytes)* and animals *(xerocoles)* show adaptations for three challenges: getting moisture, conserving moisture, and defending oneself from biotic and abiotic factors. Many of these adaptations are just physiological and behavioral:

- Different plants and animals adopt different mechanisms to procure enough water. The sand lizard *(Molcoh)* and horned toad *(Phrynosoma)* have hygroscopic skin to absorb moisture like the blotting paper even from unsaturated air. The kangaroo rat *(Dipodomys)* fulfills its water needs from metabolic synthesis. Others satisfy their water needs through the food they consume.

- Desert animals prevent water loss from their body by reducing surface area, making skin impermeable through its thickening and hardening, as well as through the presence of scales and spines *(Phrynosoma, Moloch)*, reducing the number of sweat glands in mammals, avoiding day heat by seeking the shadows of rocks and becoming active at night *(nocturnal)*, and excreting wastes as solid dry pellets.

- Some desert animals store water in their body and use it economically; the camel stores water in the tissues all over the body, whereas the desert lizard *(Uromastix)* stores it in the large intestine.

- Because of sand and dust in the air, the ears, eyes, and nostrils are protected by valves, scales, fringes, eyelids, or by being reduced in size.

- Jackrabbits *(Lepus)*, [fox]es *(Vulpes velox)*, others have large pinnae to function as efficient heat radiators without having to lose moisture.

- Coloration and behavior allow animals to harmonize with the desert surroundings. For example, sand colored and rough skinned *Phrynosoma* on detecting threats digs in the sand to obliterate the body contour and to harmonize in the background.

- Possession of venom (poison) is for self–defense and almost all desert snakes and spiders are poisonous.

Protective Adaptation

Protection from enemies, predators, and even mistakes is achieved by the use of protective devices and mechanisms, such as slippery surfaces, horns, spines, unpleas-ant smells (e.g., shrew), poison, hard shells, *autotomy* (self-cutting) of tail (e.g., wall lizard), or by the use of coloration together with behavioral postures. Colorations are used for different purposes:

- Cryptic coloration or camouflage is for making the animals invisible or indistinct from the environment by assimilating with the background or by breaking up the body contour. Animals living in snowy conditions may be white, forest animals may be striped or spotted, and desert animals may be sandy colored. The *chameleon* has several layers and varieties of chromatophores that enable it to change its colors according to the color of the surroundings.

- Resemblance coloration, together with morphological features and behavioral postures, make the animals resemble exactly the particular uninteresting ob-jects of the environment, thus deriving protection. Some of the examples are stick insects, leaf insects *(Phyllium)*, and others.

- Warning coloration is meant to avoid the mistake encounter of dangerous animals in general, or the encounter of unpalatable organisms by predators. The animals bear this coloration to advertise their being dangerous or unpalatable. Gila monster *(Heloderma)*, the only known poisonous lizard, has bright black, brown yellow and orange bands. Most poisonous snakes possess warning color-ation. Bees and wasps warn others of their stings.

- Mimicry is defined as the imitation of one organism by another for the purpose of concealment, protection, or other advantages. The species that imitates is called a *mimic* and the one which is copied a *model*. Depending on the purposes of mimicry, it can be protective or aggressive:

  ○ Protective mimicry is a protective simulation by a harmless species in form, appearance, color, and behavior of another species that is unpalatable or dangerous. For example, certain harmless flies with a pair of wings may mimic four winged bees or wasps that are well known dangerous insects, thus deriving protection. This is *Batesian mimicry*. If two species have same warning coloration and mutually advertise their dangerousness or unpalatability so as to make predators learn to avoid both of them, then it is called *Mullerian mimicry*.

  ○ Aggressive mimicry is used by predators. Here, a predator mimics the organism favored by its prey so as to trap the latter. For example, the African lizard resembles a flower, or a spider may resemble the flower of an orchid, and so forth.

Volant Adaptation

Volant adaptation refers to adaptations in those having a flying mode of life. Included are modifications that help organisms sustain and propel their body in the air. It may be for passive gliding or for active true flight.

Passive Gliding

These types of movements involve no propulsion other than the initial force of jumping and gravitational force. It is characterized by leaping or jumping from a high point and being held up by some sustaining organs to glide to the lower levels.

- The skin on either side of the body become expanded and stretched between fore and hind limbs to form what is called *patagium*. Patagia are sustaining organs in many animals, including the flying squirrel *(Sciuropterus)* and flying lemur *(Galeopithecus volans)*. In the flying lizard *(Draco)*, the patagia are supported by 5/6 elongated ribs.

- The flying frog *(Rhacophorus)* possesses very large webbed feet for sustaining purposes. Its digits terminate in adhesive pad to ensure clinging on the landing surface.

- In flying fish *(Exocoetus)*, the pectoral fins are enlarged to form gliding surfaces and the ventral lobe of the caudal fin is elongated to make dashes on the water surface to push the animal for the gliding flight. The fish makes this flight for 200 to 300 meters to escape from large fish. Other genera of flying fishes are *Dactylpterus*, *Pantodon*, and *Pegasus*.

Active True Flight

Active true flight is aerial flight with both sustaining and propulsion; it is found among living forms in insects, birds, and bats. Being widely different groups, it is held that their flight developed independently. Nonetheless, they show many common features:

- Though the flight organs in all the groups are wings, their structure varies greatly:

  - Insect wings are made up of cuticle strengthened by thickening called veins. Typically, there are two pairs of wings developed on the dorso–lateral sides of the meso– and meta–thoracic segments. In Diptera, only meso–thoracic wings are developed.

  - Bat wings are modified forelimbs. The humerus is well developed and the radius is long and curved, while the ulna is vestigial. The pollex (thumb) is free and clawed for crawling and climbing. The patagia are supported by elongated second, third, fourth, and fifth digits.

  - Bird wings are also the modification of forelimbs, but with reduced digits. They represent the most specialized wings among the modern wings. The feathers of flight are borne on the arm and hand, forming well expanded wings.

- Sternum (breast bone) is well developed for the attachment of flight muscles. In bird, it is keeled.

- Specifically strong flight muscles are present.

- Body is made light especially in birds due to the:

    ◦ presence of pneumatic bones.

    ◦ reduction of internal organs, e.g., ovary and oviduct of right side, urinary bladder.

    ◦ presence of air sacs in the body.

    ◦ presence of light feathers covering the body.

- Especially in birds, the optic lobe of the brain is highly developed, correlating with which eyes are also large to ensure good sense of vision. To overcome sudden change in air pressure, the eyes bear characteristic sclerotic plates and also comb–like, vascular, and pigmented structures called pectin. They regulate the fluid pressure within the eyes.

- The conversion of forelimbs into wings in birds is compensated by the presence of toothless horny beaks and long flexible necks.

The Theories of Adaptation

Jean-Baptiste Lamarck was among the first to put forth a theory of adaptation, offering a process by which such adaptations could have arisen. His theory was referred to as the inheritance of acquired characters. But it failed to explain the origin and inheritance of characters as a population phenomenon. Epigenetics (Pray 2004) and Baldwinian evolution offer analogous processes in modern evolutionary theory.

Jean-Baptiste Lamarck

Next, Charles Darwin came up with a more concrete explanation of adaptation that fit with observations. His theory of natural selection offered a mechanism by which suitable characters for particular environments could come to gradually predominate in the polymorphic population. So popular is Darwinian Theory that the term adaptation is sometimes used as a synonym for natural selection, or as part of the definition ("Adaptation is the process by which animals or plants, through natural selection, come to better fit their environment.") However, most biologists discourage this usage, which also yields circular reasoning. Nonetheless, Darwin's theory does not give reasons for the underlying polymorphism on which natural selection works, and evidence of natural selection being the directing force of changes on the macroevolutionary level, such as new designs, is limited to extrapolation from changes on the micro evolutionary level (within the level of species).

Industrial melanism is often presented as the best illustrative example of evolution of adaptive modification. In this case, two forms of peppered moths *(Biston betularia)* exist, melanic and non-melanic forms. Field studies in England over a 50-year period suggest that melanic forms increased in proportion in polluted areas because of the phenomenon of industrial melanism. This shift toward darker melanic forms is attributed to a heightened predation by birds of the light-colored moths, because the lighter forms could more easily be seen on the tree trunks that have been increasingly darkened from pollution. However, Wells pointed out that there are flaws in the studies, including the fact that peppered moths do not normally alight on tree trunks, and there are even inverse correlations with pollution in many situations.

Coadaptation

Coadaptation is the correlation of structural or behavioral characteristics in two or more interacting organisms in a community or organs in an organism resulting from progressive accommodation by natural selection.

Coevolution and coadaptation are an integral part of the biological evolution of plants, animals and microorganisms that live together in the same ecosystem. Coevolution and coadaptation are a game of mutual adjustment and change that never ends.

The same thing happens between humans and the rest of the ecosystem. Human social systems adapt to their environment, the ecosystem, and ecosystems adapt to human social systems. Natural ecosystems, and the natural parts of agricultural and urban ecosystems, respond to human interventions by making adjustments that promote survival. Agricultural and urban ecosystems also evolve and adapt to the social system as people change them to fit with their changing society.

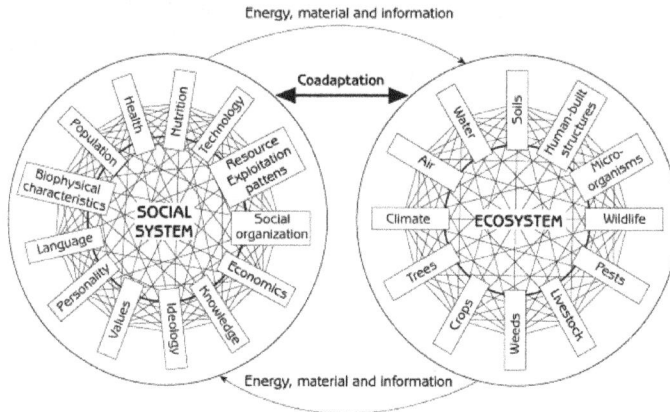

Figure: Interaction, coevolution and coadaptation of the human social system with the ecosystem

Examples of Coadaptation

Coadaptation of People and Mosquitoes

Approximately 100 years ago, the French moved large numbers of people in colonial Vietnam from the lowlands to the mountains. They wanted more people in the mountains to cut forests, work on rubber plantations and work at tin mines. Unfortunately, many lowland people died of malaria when they were forced to live in the mountains. This was surprising, because malaria had not been a serious problem in Vietnam. Malaria is transmitted by mosquitoes but, fortunately for lowland people, the species of mosquito that breed in the vast rice fields of the lowlands do not transmit malaria. Although the mountains have malaria-transmitting mosquito species, the disease was never a serious problem for the mountain people, who lived there for many generations. Because of malaria, the French never succeeded at moving large numbers of lowland people to the mountains.

Why did lowland people get malaria when mountain people did not? The reason was a difference in culture. Mountain people build their houses raised above the ground, keep their animals such as water buffalo below the house and have their cooking fire inside the house. Mosquitoes fly close to the ground, prefer to bite animals instead of people and are repelled by smoke, so they seldom enter the raised, smoke-containing houses of the mountain people, and bite the animals beneath the houses instead of the people.

Lowland people build their houses right on the ground, keep their animals away from the house and cook outside. When lowland people moved to the mountains, they continued to build their houses and cook in the traditional way. Mosquitoes easily entered the ground-level, smoke-free houses and bit the people within the houses because there were no animals to attract them away. The lowland house design worked well in the lowlands but was not adapted to the mountain ecosystem.

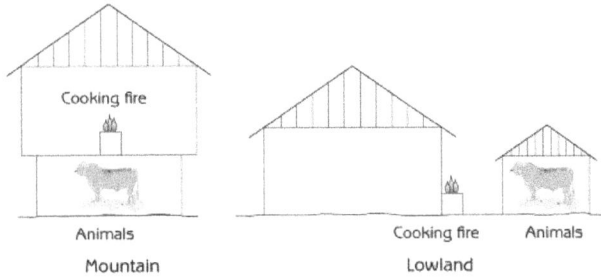

Figure: Traditional house design of mountain people and lowland people in Vietnam

The mountain people were protected from malaria without realizing that mosquitoes transmit the disease. At that time, before scientists discovered the role of mosquitoes in malaria transmission, people everywhere in the world believed that malaria was caused by spirits or contaminated water. If mountain people were asked why they built their houses in a specific way, they would say it is tradition. Their house design was a product of centuries of cultural evolution that adapted their buildings to all of their needs, including health.

In 1940, scientists invented DDT, an effective insecticide against mosquitoes that transmit malaria. Because malarial mosquitoes rest on house walls, and because DDT stays on surfaces for months after it is first applied, it was possible to kill almost all of the mosquitoes by spraying house walls with DDT just a few times a year. The World Health Organization mounted a global DDT campaign against malaria in the 1950s, and at first it worked perfectly. Malaria almost disappeared by the end of the 1960s. However, the mosquitoes came back during the 1970s, and so did malaria. About 500 million people now suffer from malaria worldwide each year, and several million of them die.

Mosquitoes returned because they evolved a resistance to DDT. A few mosquitoes had a genetic mutation that protected them from DDT. After DDT went into heavy use, this DDT-resistant gene spread quickly through the mosquito populations because mosquitoes with the resistant gene survived when other mosquitoes were killed. There was also a behavioral mutation in some regions. The mosquitoes started to rest on vegetation outside of houses instead of on house walls sprayed with DDT. DDT was not a sustainable technology for malaria control. But what about other insecticides? DDT is very inexpensive; however, all other insecticides are too costly for large-scale use against malarial mosquitoes. Most countries gave up on controlling the disease, and there has been little progress in controlling malaria since then. The use of anti-malarial drugs has reduced fatalities in some areas, but many of these drugs no longer work because the malarial parasite has evolved a resistance to them.

Controlled Burning by Native Americans

This is something that Native Americans knew long before Europeans came to North America. Because Native Americans coevolved with North American ecosystems for thousands of years, their social system and their technology for using the land were highly adapted to a sustainable relationship with the environment. Controlled burning was an integral part of their forest management. They started fires because they knew that frequent fires were a way to maintain healthy forest ecosystems. They also used controlled burning to create small patches of other ecosystems, such as grassy meadows. A landscape mosaic comprising different stages of ecological succession provided more wild plants and animals as food sources than a landscape with only forest. When Europeans came to North America, they made numerous mistakes because their social system was not adapted to North American ecosystems.

Coevolution of the Social System and Ecosystem from Traditional to Modern Agriculture Ecosystems adapt to human social systems in two ways:

1. Ecosystems reorganize themselves in response to human actions.

2. People change ecosystems to fit their social system.

The mosquito example illustrated how natural ecosystems reorganize themselves. Mosquitoes evolved DDT resistance in response to a high death rate imposed by DDT. The natural components of agricultural and urban ecosystems also adapt to human actions by reorganizing themselves. The parts of agricultural and urban ecosystems that are organized by people change with the social system because people change them. People make agricultural and urban ecosystems to fit their social system, and people adjust their social system to fit with their agricultural and urban ecosystems. The modernization of agriculture after the Industrial Revolution illustrates coevolution of the social system with agricultural ecosystems.

Before the Industrial Revolution, people were very much aware of environmental limitations. Their culture, values, knowledge, technology, social organization and other parts of their social system were by necessity closely adapted to nature. Most people were small-scale subsistence farmers; most of the agricultural production was for home consumption. Most families had a variety of farm animals and cultivated many different crops to meet the family's needs for food and clothing. Agricultural techniques were adapted to local environmental conditions. The amount of land that each family could cultivate was limited by the large amount of human or animal labor that was necessary for agriculture. Most farmers used polyculture - a mixture of several crops together in the same field.

Polyculture had a number of advantages:

*   It protects the soil from erosion, and can maintain soil fertility without the use of chemical fertilizers. The mixture of different crop species in a polyculture creates a large quantity of vegetation, which covers the ground completely. In contrast, it is common for a monoculture (one crop) field to have a lot of bare ground. The large quantity of vegetation in polyculture protects the soil from falling rain, thereby reducing soil erosion. The vegetation also provides substantial amounts

of organic fertilizer when the unused part of the crop is ploughed back into the soil. If some of the plants in the field are legumes (for example, beans or peas), bacteria in the roots of the legumes convert atmospheric nitrogen to forms that plants can use.

- It provides natural pest control. Agricultural pests are usually specific to a particular kind of crop. For example, if a field is 100 per cent corn monoculture, corn pests multiply to large numbers and inflict a lot of damage if pesticides are not used. However, if a field has many different crops, with only a few corn plants, the corn pests have trouble finding their host; as a result, they are unable to multiply to large numbers and the damage is limited. Polyculture also provides good habitat for animals such as birds and predatory insects that eat insect pests. Predators provide natural control of the pests. When chemical pesticides are used in modern agriculture, many of the predators are killed, and much of the natural pest control is lost.

- It allows farmers to diversify their risks. If the weather during one year is bad for some kinds of crops, it will probably not be bad for all crops. If market prices are low for some crops, the prices will probably be better for other crops.

Agriculture changed in Europe when the Industrial Revolution made it possible to use machines instead of human and animal labor for work such as ploughing fields and harvesting crops. Machines gave farmers the ability to cultivate larger areas of land. Farm sizes increased dramatically because mechanized agriculture is more efficient on a larger scale (economy of scale). These initial changes in the social system and the ecosystem set in motion a series of changes through interconnected positive feedback loops in the ecosystem and social system.

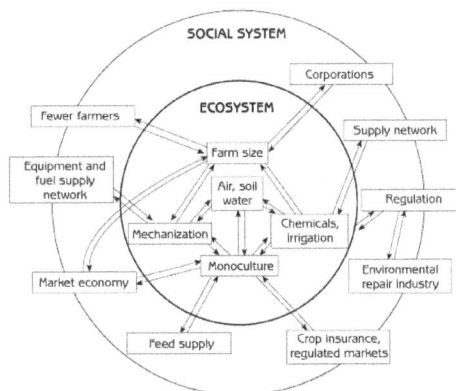

Figure: Interaction of the social system with agricultural ecosystems after the Industrial Revolution

When farm sizes increased, farmers were able to produce more than they needed for their own families, so they changed from subsistence farming to a market economy. Larger farm size also meant that there was surplus production to support cities. Many people got out of farming and moved to cities, where economic opportunities were better.

One of the main changes in the ecosystem was from polyculture agriculture to monoculture. With mechanization, farmers stopped mixing crops together because farm machines work best with single crops. The market economy also provided an incentive to change from polyculture to monoculture because producing and marketing a single crop was more convenient for farmers. The change from polyculture to monoculture led to many other changes. Monoculture did not protect the soil from erosion or maintain soil fertility as well as polyculture did. Risks of crop failure due to bad weather or pest attacks were also greater with monoculture because 'all the eggs were in one basket'. As a result, it was important to make agriculture more independent of the environment by means of irrigation, chemical fertilizers and pesticides - all of which were possible with new developments in science and energy from fossil fuels.

Governments gradually became involved in research to provide better technology for the new style of agriculture: improved crop varieties to provide higher yields with high inputs (chemical fertilizers, pesticides, etc.), as well as better techniques for using the inputs. Commercial networks were set up to provide machines, chemicals and high-yield crop seeds to farmers. Government crop insurance and market regulation, including government subsidies, were developed because of the higher risks associated with monocultures.

The improvements in technology made monoculture even more advantageous compared to polyculture. Because different plants have different growth requirements, conditions in a polyculture field cannot be optimal for all of the different species of plants that grow together. Specialization through monoculture made it easier for a farmer to use high inputs to provide optimal conditions to attain the highest yields with one particular crop.

Farmers changed their belief system - their worldview. Once the Industrial Revolution was underway, technology, machines and fossil fuels seemed to free people from many environmental limitations. People began to think of agriculture more in economic terms, as a business enterprise, and less in environmental terms. Everyone believed that the future was going to provide advances in science and technology that offered endless possibilities for capital accumulation and economic development. Eventually, many farms were taken over by large corporations, and agriculture became more and more 'vertically integrated'. Today many of the same corporations that own supermarkets also own the farms and food processing factories that supply their supermarkets with food.

People eventually had to make changes to their beliefs as the environmental and human health implications of pesticides and chemical fertilizers became apparent in recent years. Governments began to regulate the use of chemicals, and they conducted research on how to deal with the consequences of applying chemicals. A new environmental industry arose in the private sector to deal with pollution from agriculture and other sources.

Throughout history, social systems and agricultural ecosystems have changed in ways that have allowed them to continue functioning well together. The same is generally true today. Modern social systems and agricultural ecosystems continue to change together and they are strongly co-adapted. The problem today is that modern agricultural ecosystems have lost their coadaptation with the natural ecosystems that surround them - natural ecosystems upon which the agricultural ecosystems depend for their long-term viability. Modern agricultural ecosystems rely on large-scale fertilizer and pesticide inputs from natural sources that may not be sustainable on such a scale, and they pollute surrounding ecosystems with fertilizer and pesticide runoff from fields. Modern agricultural ecosystems also depend upon natural ecosystems for massive inputs of energy and, in many instances, irrigation water, which may not be possible to sustain.

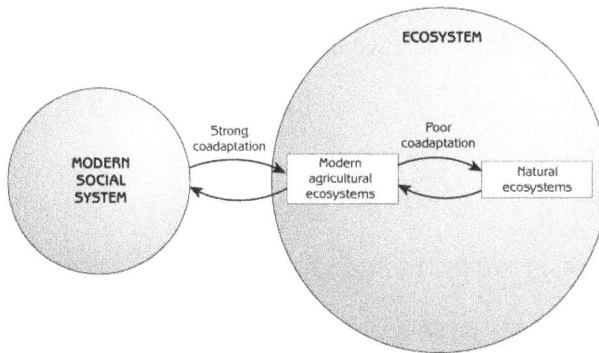

Figure: Coadaptation of modern social systems and ecosystems

The recent popularity of organically grown foods is stimulating a return to agricultural ecosystems that are more compatible with natural ecosystems. Organic farmers are returning selectively to traditional farming methods while employing organic fertilizers and environmentally benign methods of pest control. Their agricultural ecosystems are not dependent on chemical inputs, and the pollution of surrounding ecosystems is minimal. As the market for organically grown foods continues to expand, agricultural scientists and farmers will be stimulated to develop new and ecologically sound agricultural technologies.

Adaptive Mutation

Adaptive mutation is defined as a process that, during nonlethal selections, produces mutations that relieve the selective pressure whether or not other, non-selected mutations are also produced. Examples of adaptive mutation or related phenomena have been reported in bacteria and yeast but not yet outside of microorganisms. A decade of research on adaptive mutation has revealed mechanisms that may increase mutation rates under adverse conditions.

Examples of adaptive mutation or related phenomena have been reported in the bacteria *Escherichia coli*, Salmonella typhimurium, Bacillus subtilis, Pseudomonas sp. and *Clostridium* sp., and, in the eukaryotic microbial species, *Saccharomyces cerevisiae* and *Candida albicans*. Whether the phenomenon exists outside of microorganisms is not known, but it seems unlikely to contribute to the evolution of organisms that have their germ lines protected from the environment. Mutations occurring in the non-dividing somatic cells of higher organisms, however, could give rise to disorders such as cancer.

Evolutionary Significance of Recombination-induced Mutation

The recombination-dependent mutational mechanism is particularly active on F, but can be expected to occur whenever a nick is encountered during DNA replication. As discussed above, accumulating evidence indicates that:

(1) Recombination is constantly active in both proliferating and non-proliferating cells,

(2) That recombination initiates DNA synthesis,

(3) That this synthesis can lead to mutations.

Thus, recombination not only rearranges existing alleles, it can also create new genetic variants via associated DNA synthesis. This may not be a major source of variation in growing organisms when other mutational mechanisms are active, but might become significant in static populations. In *E. coli*, many of the components of this system—the recombinase, the specialized polymerases and the enzymes that help resolve recombination intermediates—are induced as part of the SOS response to DNA damage. SOS genes are also induced in aging colonies and at the end of growth in complex medium. Similar inducible systems may exist in other organisms, allowing limited transient mutator states to be active when genetic variability may be advantageous.

Strain FC40 detects frameshift mutations, and it might be argued that these are not particularly significant in evolution (although they are important in switching certain antigen-determinant genes on and off). The mutagenic mechanism, however, is not specific for frame shifts; all types of mutation may be produced, particularly when the error-prone polymerases are induced. I hypothesize that both of *E. coli*'s error-prone polymerases can participate in double-strand end repair, but since Pol IV is more readily induced than Pol V it dominates the response. Although Pol IV tends to make frame shifts, both Pol IV and Pol V also make base substitutions. Base substitutions are considered to be more evolutionarily significant than frame shifts because these mutations are more likely to alter, not eliminate, protein function.

References

- Carroll SB, Grenier JK, Weatherbee SD (2005). From DNA to Diversity: Molecular Genetics and the Evolution of Animal Design (2nd ed.). Malden, MA: Blackwell Publishing. ISBN 978-1-4051-1950-4. LCCN 2003027991. OCLC 53972564

- Mutation-genetics, science: britannica.com, Retrieved 25 March 2018

- Orengo CA, Thornton JM (July 2005). «Protein families and their evolution-a structural perspective». Annual Review of Biochemistry. 74: 867–900. doi:10.1146/annurev.biochem.74.082803.133029. PMID 15954844

- Neutrality-and-molecular-clocks-100492542: nature.com, Retrieved 08 June 2018

- Montelone BA (1998). "Mutation, Mutagens, and DNA Repair". www-personal.ksu.edu. Archived from the original on 26 September 2015. Retrieved 2 October 2015

- Point-mutation, science: britannica.com, Retrieved 29 April 2018

- Freese E (June 1959). "The specific mutagenic effect of base analogues on Phage T4". Journal of Molecular Biology. 1 (2): 87–105. doi:10.1016/S0022-2836(59)80038-3

- Frameshift-mutation-frame-shift-mutation-frameshift-203: nature.com, Retrieved 17 May 2018

- Bohidar HB (January 2015). Fundamentals of Polymer Physics and Molecular Biophysics. Cambridge University Press. ISBN 978-1-316-09302-3

- Polar-mutation, molecular-biology-glossary-10974: genscript.com, Retrieved 17 July 2018

- Long M, Betrán E, Thornton K, Wang W (November 2003). "The origin of new genes: glimpses from the young and old". Nature Reviews Genetics. 4 (11): 865–75. doi:10.1038/nrg1204. PMID 14634634

Permissions

Index

www.ingramcontent.com/pod-product-compliance
Lightning Source LLC
Chambersburg PA
CBHW061953190326
41458CB00009B/2864